Textbook of

Physical Chemistry

Volume I

Textbook of
Physical Chemistry
Volume I

N.K. Vishnoi
Professor,
Department of Basic Sciences
N.I.I.T. University
Neemrana (Rajasthan)

&

R.J. Shukla
Ex. Head Deptt. of Chemistry,
Kamla Nehru Institute of Physical & Social Sciences,
Sultanpur (U.P.)

Ane Books Pvt. Ltd.
New Delhi ♦ Chennai

Textbook of Physical Chemistry Vol. I

N.K. Vishnoi & R.J. Shukla

Published by

Ane Books Pvt. Ltd.

4821, Parwana Bhawan, 1st Floor, 24 Ansari Road,
Darya Ganj, **New Delhi** - 110 002, India
Tel.: +91(011) 23276843-44, Fax: +91(011) 23276863
e-mail: kapoor@anebooks.com, Website: www.anebooks.com

Branch Office :

- Avantika Niwas, 1st Floor, 19 Doraiswamy Road, T. Nagar,
  Chennai - 600 017, Tel.: +91(044) 28141554, 28141209
  e-mail: anebooks_tn@airtelmail.in

ISBN : 978-81-90832-26-7

Printed at : R.K. Offset Process, New Delhi.

Preface

Physical chemistry is the branch of chemistry that deals with the principles of physics underlying chemical interactions and transformation of material in order to correlate the quantitative as well as a qualitative aspect of matter energy interactions. A course in physical chemistry forms an essential part of the curriculum of not only science students but also of students in various branches of engineering and technology. It therefore becomes important to define the scope of the book on physical chemistry. Presently we have chosen the model syllabus suggested by panel of experts appointed by the university grants commission in the quest for uniformity of syllabi in the Indian universities.

As per the U.G.C. course structure, the entire study material of physical chemistry at graduate level has been divided into three units with each unit including the course material expected to be studied in each of the three years of a B.Sc. course. Volume one of the present book of physical chemistry deals with unit I of the above course which includes states of the matter–gaseous, liquid, solid and colloidal state and chemical kinetics along with catalysis.

Any course in physical chemistry incorporates a large number of definitions and concepts which need to be explained to the students through application of these in the from of solved exercises as well as numerical problems. Therefore in order to develop the understanding of the basic principles and fundamentals of the subject we have introduced solved example which are interspersed throughout the text. To sustain the interest of the student in the text a number of examples have been incorporated in the text. This is followed by objective, short answer and subjective questions. It has been our attempt to ensure the incorporation of essentials of the subject without compromising on the accuracy of the concepts. Unified treatment of the subject and explanation of facts at molecular level are the highlights of the present text.

In addition to the above there is an introductory unit which includes chapters on 'Mathematics for Chemists' and 'Computer for Chemists' as required in the U.G.C. model syllabus. We sincerely hope that the present effort with help the students in acquiring a better understanding of the subject.

We wish to thank earlier authors of the subject, teachers and students who have all contributed in shaping our thoughts while writing the present text. Thanks are due to our Publisher M/S Ane Books Pvt. Ltd. for bringing out the books in an excellent form and to Dr. S.N. Pandey for encouraging us to write the present text.

Constructive criticism and suggestions for the improvement of the text are Welcome.

N.K. Vishnoi

R.J. Shukla

Contents

2. Computers for Chemists

(Pages 27-38)

UNIT-II

1. Gaseous State

(Pages 41-70)

2. Liquid State

(Pages 71-84)

3. Solid State

(Pages 85-104)

4. Colloidal State

(Pages 105-120)

5. Chemical Kinetics and Catalysis

(Pages 121-174)

UNIT-I

1

Mathematical Concepts

Most of the present studies undertaken in chemistry, particularly physical chemistry, require some basic understanding of simple modern mathematical operations and tools. For example, study of chemical kinetics and quantum chemistry requires a knowledge of differential and integral calculus. Use of logarithm and curve sketching is also a necessity in chemistry. It is, therefore, necessary that these topics of mathematics are introduced to a learner of physical chemistry.

1.1 LOGARITHM

1.1.1 Definition of Logarithm

Consider an exponential relation, say,

$$a^x = y \tag{1.1}$$

where, 'x' is any rational number and 'a' is a positive real number (but not 1).

Here, logarithm of 'y' to the base of 'a' is x which is expressed as:

$$\log_a y = x \tag{1.2}$$

■ **Example:** The expression $10^2 = 100$ may be written as $\log_{10} 100 = 2$, which implies that logarithm of 100 to the base 10 is equal to 2.

Important Points to Remember

(*i*) The logarithm of 1 to any base is zero, *i.e.*, $\log_0 1 = 0$

(*ii*) The logarithm of any number to the same base is always 1. Thus, $\log_3 3 = 1$ and $\log_{100} 100 = 1$

Problem 1. Find the value of x in the expression $\log_x \frac{1}{10} = -1$.

Solution: Writing the given expression in the form $a^n = y$, we have

$$x^{-1} = \frac{1}{10} = 10^{-1}$$

$$\Rightarrow \quad x = 10$$

1.1.2 Some Important Laws of Logarithm

(*i*) $\log_e (m \times n) = \log_e m + \log_e n$ (1.3)

(*ii*) $\log_e (m/n) = \log_e m - \log_e n$ (1.4)

(*iii*) $\log_a (m)^n = n \log_a m$ (1.5)

(*iv*) $\log_a m \times \log_n a = \log_n m$ (1.6)

(*v*) $\log_y x \times \log_x y = 1$ (1.7)

1.1.3 Common Logarithm

The logarithm of a number to the base of 10 is called common logarithm.

1.1.4 Characteristic and Mantissa of Logarithm

The value of logarithm of a number consists of

(*a*) **Characteristic:** It is an integer which may be positive, negative or zero. Characteristic of the logarithm of a single digit number before decimal is zero; that of a number containing more than one digit is positive and equal to one digit less than the total number of digits and written before the decimal point. The negative characteristic is indicated by putting a bar on the concerned digit. For example, Characteristic of $\log_{10}$ 1000.1021 is 3, $\log_{10}$ 100.1021 is 2, $\log_{10}$ 1.1021 is 0, $\log_{10}$ 0.1021 is – 1 or $\bar{1}$ and $\log_{10}$ 0.01021 is $\bar{2}$.

(*b*) **Mantissa:** It is the decimal part of logarithm, obtained from logarithm table. A part of this table is presented below :

	0	1	2	3	4	5	6	7	8	9	1	2	3	4	5	6	7	8	9
14	1461	1492	1523	1553	1584	1614	1644	1673	1703	1732	3	6	9	12	15	17	20	23	26
15	1761	1790	1818	1847	1875	1903	1931	1959	1987	2014	3	6	8	11	14	17	19	22	25

Thus,

Mantissa of $\log_{10}$ 14 = 0.1461 (the value of number 14 under column 0)

Mantissa of $\log_{10}$ 145 = 0.1614 (the value of number 14 under column 5)

Mantissa of $\log_{10}$ 1455 = 0.1614 (the value of number 14 under column 5 +

+ 0.0015 the value of mean

= 0.1629 difference under column 5)

Combining (*a*) Characteristic and (*b*) Mantissa, the total value of logarithm of 1455

= Characteristic + Mantissa

= 3 + 0.1629

= 3.1629

1.2 ANTILOGARITHM

Logarithm and antilogarithm are two mutually reverse operations.

We know that $\log_{10} 1000 = 3$

So, antilog of 3 = 1000

Antilog is found from the antilogarithm table, taking only the decimal part of the given number. Once the antilog of decimal part is obtained, decimal point is inserted using the following procedure:

1. For positive characteristic value, say m, the decimal point is inserted such that first significant figure appears at $(m + 1)$th place.

   For example, the antilog of

$$0.1135 = 1297 \quad \underline{+2} \quad = 1.299$$

[decimal point being inserted at $(m + 1)^{th}$ place]

Antilog of $1.1135 = 12.99$

Antilog of $2.1135 = 129.9$

Antilog of $\bar{1}.1135 = 0.1299$ (the first significant figure being at 1st place)

Antilog of $\bar{2}.1135 = 0.01299$ (the first significant figure being at 2nd place)

Problem 2. If $\log_{10} x = 2.3617$, find the value of x.

Solution: Here, x = antilog 2.3617

(from antilog table without decimal place)

$$= 2296 + 0004$$
$$= 2300$$

On inserting decimal point

$$x = 230.0$$

Problem 3. If $\log_{10} x = -2.5453$, find the value of x.

Solution: Given that $\log_{10} x = -2.5453$

$$= -3 + 3 - 2.5453$$
$$= \bar{3}.4547$$

$\therefore$ $x = \text{antilog } \bar{3}.4547$

$$= 2844 + 0005$$
$$= 2849 \quad \text{(without decimal place)}$$

On inserting decimal point,

$$x = 0.002849$$

1.3 CURVE SKETCHING

1.3.1 Introduction

Curve sketching is the graphical representation of data related to a given equation of dependent and independent variables.

Following are the various forms of curves:

(*i*) Cartesian Equations: These equations are the relations between cartesian co-ordinates x and y, *e.g.*, $y = 4x$, $x^2 + 2xy + y^2 = 0$, *etc.* (1.8)

(*ii*) Polar Equations: These equations co-relate polar coordinates r and θ, *e.g.*,

$$r = 2 \sin \theta \tag{1.9}$$

(*iii*) Parametric Equations: These equations are relations of x and y co-ordinates with a third parameter, *e.g.*,

$$x = a \sin t$$

1.3.2 Tracing of Cartesian Curves

It is based on the following points:

(*a*) Symmetry: The symmetry of the curve is determined by using certain laws stated as under:

(*i*) If there are only even powers of y in an equation, the curve will be symmetrical about x-axis, *e.g.*, the curve traced for equation $y^2 = 4ax$ will be lying uniformly on both sides of x.

(*ii*) When the equation contains only even powers of x, such as $x^2 = 4ay$, the curve traced using this equation will be symmetrical about y-axis. In view of points (*i*) and (*ii*), the curve traced for equation $x^2 + y^2 = a^2$ would be symmetrical about both the axes.

(*iii*) The curve would be symmetrical about the line $y = x$, if the equation representing the curve remains unaltered on interchanging x and y, *e.g.*, by interchanging x and y in equation $x^2 + y^2 = a^2$, we get the equation $y^2 + x^2 = a^2$, which is the same equation and thus the curve would be symmetrical about the line $y = x$.

(*iv*) If the equation representing the curve remains unaltered on changing the sign of x and y, such a curve would be symmetrical in opposite quadrants, *e.g.*, in equation $x^3 + y^3 = 3ay^2 x$, if we interchange the sign of x and y, we get

$$(-x)^3 + (-y)^3 = 3a(-y)^2(-x)$$

or

$$x^3 + y^3 = 3ay^2x \tag{1.10}$$

which is the same equation. Hence, the curve obtained using this equation would be symmetrical in opposite quadrants.

(*b*) Points: The points where the curve intersects axes can be ascertained as:

(*i*) By putting $y = 0$ in the equation and solving for values of x, we get the points where the curve intersects x-axis, *e.g.*, in equation

$$4y^2 = x^3 (4 - x)$$

if we puts $y = 0$, we get $x = 4$ and 0.

Thus, this curve intersects x-axis at $x = 0$, $y = 0$, *i.e.*, the origin and at $x = 4$, $y = 0$.

(*ii*) When we substitute $x = 0$ in the above equation of the curve and solve for y, the curve cuts at the point $x = 0$, $y = 0$, *i.e.*, the origin.

(*c*) **Shape at the Origin:** Once the curve passes through the origin, its shape at the origin could be determined using the following conditions:

(*i*) If there is a node.

(*ii*) If there is a cusp.

(*iii*) If there is a conjugate point.

For this, we have to first write the equation of tangents at the origin by equating the lowest degree terms to zero. Then,

(*i*) There will be two real and non-coincident tangents for node.

(*ii*) There will be two real and co-incident tangents for cusps.

(*iii*) There will be two imaginary tangents for conjugate points.

In case only one tangent is obtained, it is indicative of either a point of inflection or the parabolic nature of curve. Further, it may either be upper or lower half of the parabola.

(*d*) **Shape of the Curve at the Points of Intersection of Axes:** When the curve intersects the axes at points other than origin, the co-ordinates of these points are determined. We can then write down equation of the curve with respects to this point by first shifting the origin to this point. It is now easier to determine its shape at the new origin by applying the rules discussed earlier.

(*e*) **Determination of Regions Where the Curve is Non-existent:** We find out such co-ordinates of x for which the values of y gets imaginary and then it is concluded that the curve is non-existent in the regions bound by the lines drawn for the values of x.

(*f*) **Asymptotes:** All real asymptotes of the curves are determined and their co-ordinates on the various axes are noted.

If the curve contains both the term x^n and y^n, it will not have asymptotes parallel to the axes. In case it does not contain the term x^n, we find the coefficient of the term x^{n-1} and equate this coefficient to zero and thus get asymptotes parallel to x-axis. In case x^{n-1} term is also absent in the equation, then we go for x^{n-2} term, and so on. Similarly, asymptotes parallel to y-axis can also be determined.

Problem 4. Trace the curve

$$y^2 (a^2 + x^2) = x^2 (a^2 - x^2)$$

Solution: The curve corresponding to the given equation can be traced on the basis of following points:

(*a*) As the curve has only even powers of x and y, it is symmetrical about both the axes.

(*b*) The curve passes through the origin, and the equation for tangents at the origin is

$$y = \pm x \tag{1.11}$$

These two tangents are real and non-coincident. Thus, we have a node at origin.

(*c*) The curve cuts y-axis at the origin and x-axis at $x = (\pm a, 0)$. Now, on transferring the origin to $(a, 0)$, the equation of the curve will be

$$y^2 [a^2 + (x + a)^2] = (x + a)^2 (-2ax - x^2) \quad (1.12)$$

and the tangent at origin will be $x = 0$, obtained by equating to zero, the lowest degree term.

(*d*) The curve does not have any real asymptote.

(*e*) Solving for y and considering only positive values as per equation,

we have the following results $y = x\left(\dfrac{a^2 - x^2}{a^2 + x^2}\right)^{1/2}$ (1.13)

(*i*) When $x = 0$, $y = 0$.

(*ii*) When x lies between 0 and a, y is real.

(*iii*) When $x = a$, $y = 0$.

(*iv*) When $x > a$, y is imaginary.

In view of the above points, shape of the curve will be as shown in Fig. 1.1.

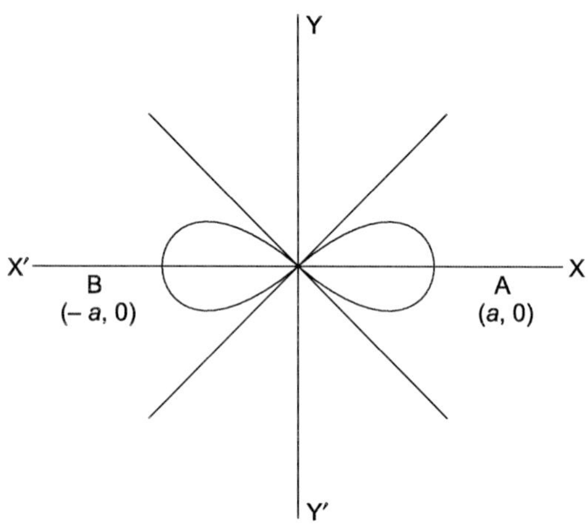

Fig. 1.1 Shape of the curve.

1.3.3 Tracing of Polar Curves

It is based on the following points:

(***a***) **Symmetry:** (*i*) If on changing the sign of θ, the equation remains unaltered, the curve is symmetrical about the initial line.

(*ii*) When only even powers of r occurs in the equation, the curve is symmetrical about the pole which, of course, is the centre.

(*iii*) If the equation remains unchanged when θ is changed to $(-\theta)$, the curve is symmetrical about the line $\theta = \pi/4$.

(*b*) **Values of *r* and θ:** (*i*) If for some real value of θ, *r* is zero, it is concluded that pole is present on the curve, and from the value of θ, equation for tangent at the pole can be derived.

(*ii*) It is appropriate to tabulate values of *r* corresponding to the values of θ ranging from 0 to 2π. A plot of such values of *r* and θ can then be drawn.

(*c*) **Asymptotes:** For a certain value of θ, when *r* has infinite length, the curve will be asymptotic in nature.

(*d*) When for certain value of θ (say between θ_1 and θ_2) the value of *r* is imaginary, the curve will not lie in such regions where $\theta = \theta_1$ and $\theta = \theta_2$.

Problem 5. Trace the curve $r = 4$.

Solution: When for every value of θ, the distance of curve from the pole remains the same, the curve will traverse a circle as shown in Fig. 1.2.

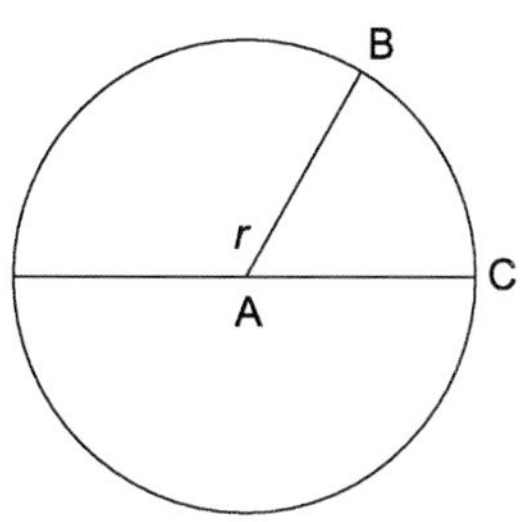

Fig. 1.2 A circle.

Problem 6. Trace the curve $r = a \sin 4\theta$.

Solution: Curve for the given equation can be traced as per following points:

(*a*) The curve is symmetrical about the line $\theta = \frac{\pi}{4}, \frac{3\pi}{4}, \frac{5\pi}{4}$

(*b*) For $r = 0$, $\theta = 0, \frac{\pi}{2}, \frac{3\pi}{4}$

Hence $\theta = 0, \frac{\pi}{2}, \frac{3\pi}{4}$ are the tangents at pole.

(*c*) The tabulated values of *r* and θ are

$$\begin{array}{lllll} \theta = 0, & \frac{\pi}{8}, & \frac{\pi}{4}, & \frac{\pi}{2}, & \frac{3\pi}{2} \\ r = 0, & r = a & r = 0 & r = 0 & r = -a \end{array}$$

Hence, the curve is as shown in Fig. 1.3.

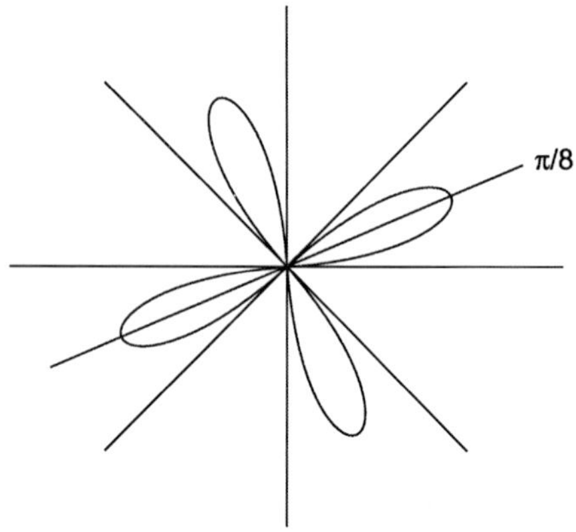

Fig. 1.3 The curve.

1.3.4 Tracing of Parametric Equations

For tracing such curves:

1. We eliminate the third parameter using the given relations, and obtain a cartesian equation of the curve.
2. We first find the values of x and y by assigning different values to the third parameter, and then find the function (dy/dx) for these values.

Problem 7. Trace the curve for

$$x = a\cos\phi$$
$$y = b\sin\phi$$

Solution: Rearranging the given equation we get

$$\cos\phi = x/a \quad (1.14)$$
$$\sin\phi = y/b \quad (1.15)$$

Now,
$$\sin\phi = \sqrt{1-\cos^2\phi} \quad (1.16)$$

$$\Rightarrow \quad \frac{y}{b} = \sqrt{1-\frac{x^2}{a^2}} = \left(\frac{a^2-x^2}{a^2}\right)^{\frac{1}{2}}$$

$$\Rightarrow \quad y = \frac{b}{a}\left(a^2-x^2\right)^{\frac{1}{2}} \quad (1.17)$$

$$\Rightarrow \quad \frac{a^2}{b^2}y^2 = (a^2-x^2)$$

$$\Rightarrow \quad a^2y^2 = b^2a^2 - b^2x^2$$

$$\Rightarrow \quad a^2y^2 + b^2x^2 = a^2b^2$$

Thus, the curve is symmetrical about both the axes.

Points where the curve cuts the axes

Putting $y = 0$ in Eq. 1.17 we have

$$0 = \frac{b}{a}\left(a^2 - x^2\right)^{\frac{1}{2}}$$

$$\Rightarrow \qquad \left(a^2 - x^2\right)^{\frac{1}{2}} = 0$$

$$\Rightarrow \qquad x = \pm\, a \tag{1.18}$$

Hence, the curve cuts x-axis at a and $-a$.

Next putting $x = 0$, we have $y^2 = b^2$

$$\Rightarrow \qquad y = \pm\, b \tag{1.19}$$

Hence, the curve cuts y-axis at b and $-b$.

Shape of the curve at points a and $-a$ on x-axis

On shifting the origin to point $(a, 0)$, Eq. 1.17 becomes

$$y = \frac{b}{a}\left[a^2 - (x-a)^2\right]^{\frac{1}{2}}$$

$$\Rightarrow \qquad = \frac{b}{a}\left[a^2 - x^2 + 2ax - a^2\right]^{\frac{1}{2}}$$

$$\Rightarrow \qquad \frac{a^2}{b^2}\,y^2 = x\,(2a - x)$$

Thus, $x = 0$ and $x = 2a$ are the equations of the tangents. (1.20)

As per above points, the curve is an ellipse with $x = a$ as the major (x-axis) and $y = b$ as the minor (y-axis) axes, as shown in Fig. 1.4.

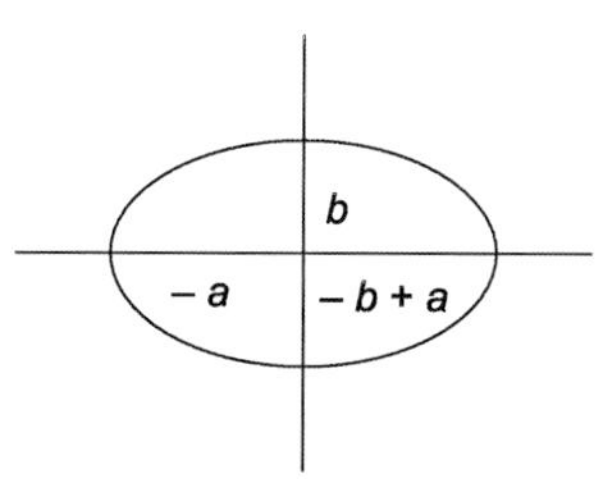

Fig. 1.4 An ellipse.

1.4 LINEAR GRAPHS AND CALCULATION OF SLOPE

Equation of a line parallel to x-axis is

$$y = b \tag{1.21}$$

Equation of a line parallel to y-axis is

$$x = a \tag{1.22}$$

Equation of x-axis is

$$y = 0 \tag{1.23}$$

Equation of y-axis is

$$x = 0 \tag{1.24}$$

Equation of a straight line ABC (Fig. 1.5) is

$$y = mx + c \tag{1.25}$$

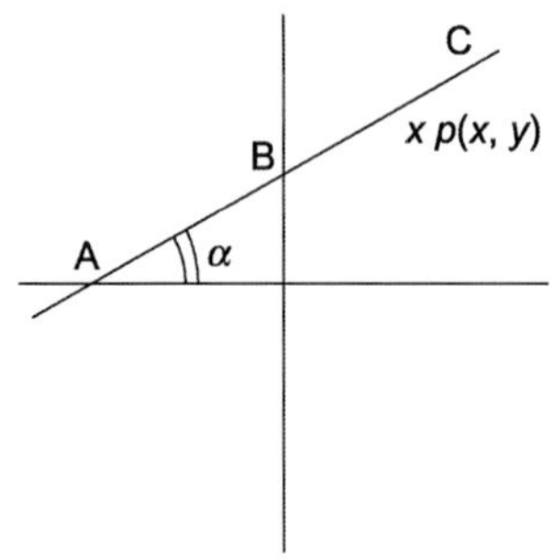

Fig. 1.5 A straight line.

Here m is called the slope of the line and c is the intercept which this line makes on y-axis. Slope is equal to the tangent of the angle made by line AC with x-axis. Let Q (x_1, y_1) be a fixed point on line AC, whose slope is m, and consider any arbitrary point P (x, y) on the line. Then, slope of the line AC is given as

$$m = \left(\frac{y - y_1}{x - x_1}\right) = \tan\alpha \tag{1.26}$$

$$\Rightarrow \quad (y - y_1) = m\,(x - x_1) \tag{1.27}$$

Equation of the line passing through any two points (x_1, y_1) and (x_2, y_2) is written as

$$(y - y_1) = \left(\frac{y_2 - y_1}{x_2 - x_1}\right)(x - x_1) \tag{1.28}$$

Problem 8. Find the equation of the line intersecting the x-axis at a distance of 4 units to the left of origin, with a slope of (– 3).

Solution: The general straight line equation is

$$(y - y_1) = m(x - x_1)$$

it passes through a point (x, y) and has a slope, m.

As the given line passes through a point (– 4, 0) and has a slope of – 3, its equation will be

$$(y - 0) = (-3)\,(x + 4)$$

$$\Rightarrow \quad y = -3x - 12$$

$$\Rightarrow \quad 3x + y + 12 = 0.$$

Problem 9. Find the slope of a line passing through points (2, 3) and (1, – 4).

Solution: *Slope m of a line passing through points is given as*

$$m : \left[\frac{y_2 - y_1}{x_2 - x_1}\right]$$

Given that (x_1, y_1) is (2, 3) and (x_2, y_2) is (1, – 4), the slope

$$(m) = \left[\frac{-4-3}{1-2}\right] = \frac{-7}{-1} = 7.$$

1.5 DIFFERENTIATION OF FUNCTIONS

1.5.1 Differential Coefficient of a Function

Consider a function defined by the equation

$$y = f(x) \tag{1.29}$$

Say, for a small change δx in x, let the corresponding change in y be δy. Then, the ratio $\left(\frac{\delta y}{\delta x}\right)$ is the average rate of change of y with respect to x in interval $(x, x + \delta x)$. When this ratio $(\delta y/\delta x)$ approaches to a certain finite value as $\delta x \to 0$, then this **limiting finite value** is called the differential coefficient of y with respect to x and denoted as

$$\left(\frac{dy}{dx}\right) = \underset{\delta x \to 0}{\text{limit}} \left(\frac{\delta y}{\delta x}\right) = \text{differential coefficient of } y \text{ with respect to } x$$

Thus, if $\quad y = f(x)$

and $\quad (y + dy) = f(x + dx)$

Then, $\quad dy = f(x + dx) - f(x)$

$$\Rightarrow \quad \left(\frac{dy}{dx}\right) = \lim_{\delta x \to 0}\left[\frac{f(x+dx) - f(x)}{\delta x}\right] \tag{1.30}$$

The process of finding the differential coefficient of a certain function of $f(x)$ is known as **differentiation**.

1.5.2 Differential Coefficient of x^n

Let $\quad y = x^n$

Then $\quad (y + \delta y) = (x + \delta x)^n$

$$\Rightarrow \quad \left(\frac{dy}{dx}\right) = \lim_{\delta x \to 0} \frac{(x+\delta x)^n - x^n}{\delta x}$$

$$= \lim_{\delta x \to 0} x^n \left[\frac{\left(1 + \frac{\delta x}{x}\right)^n - 1}{\delta x} \right]$$

$$\left(\frac{dy}{dx}\right) = \lim_{\delta x \to 0} x^n \left[1 + \frac{n\,\delta x}{x} + \frac{n(n-1)}{\underline{|2}} \left(\frac{\delta x}{x}\right)^2 + \ldots - 1 \right]$$

$$\left\{ \text{Expansion of } (1+x)^n = 1 + nx + \frac{n(n-1)x^2}{\underline{|2}} + \frac{n(n-1)(n-2)x^3}{\underline{|3}} + \ldots \right\}$$

$$= \lim_{\delta x \to 0} x^n \left[\frac{n}{x} + \frac{n(n-1)}{\underline{|2}} \left(\frac{\delta x}{x}\right) + \ldots \right]$$

containing higher powers of δx

$$= x^n \left(\frac{n}{x}\right) = n\,x^{n-1}$$

Hence $$\frac{dx^n}{dx} = n\,x^{n-1} \tag{1.31}$$

1.5.3 Differential Coefficient of e^x

Let $y = e^x$; then $(y + \delta y) = e^{(x + \delta x)}$

$$\Rightarrow \quad \left(\frac{dy}{dx}\right) = \lim_{\delta x \to 0} \left(\frac{e^{x+\delta x} - e^x}{\delta x} \right) = \lim_{\delta x \to 0} \frac{e^x \left(e^{\delta x} - 1\right)}{\delta x}$$

$$\left\{ \text{Expansion of } e^x = 1 + \frac{x}{\underline{|1}} + \frac{x^2}{\underline{|2}} + \frac{x^3}{\underline{|3}} + \ldots \right\}$$

$$\frac{dy}{dx} = \lim_{\delta x \to 0} e^x \frac{\left[1 + \delta x + \frac{(\delta x)^2}{\underline{|2}} + \frac{(\delta x)^3}{\underline{|3}} + \ldots - 1 \right]}{\delta x}$$

$$= \lim_{\delta x \to 0} e^x \left[1 + \frac{\delta x}{\underline{|2}} + \frac{(\delta x)^2}{\underline{|3}} + \ldots \right] = e^x$$

Hence, $$\frac{de^x}{dx} = e^x.$$

1.5.4 Differential Coefficient of Some Other Functions

$$\frac{d}{dx} \log_e x = \frac{1}{x}$$

$$\frac{d}{dx} \sin x = \cos x$$

$$\frac{d}{dx}\cos x = -\sin x$$

$$\frac{d}{dx}kf(x) = k\frac{d}{dx}\{f(x)\}; \text{ here, } k \text{ is a constant}$$

$$\frac{d}{dx}[f(x) \pm g(x)] = \frac{d}{dx}f(x) \pm \frac{d}{dx}g(x)$$

where, $f(x)$ and $g(x)$ are two different functions of x.

1.5.5 Differential Coefficient of a Product of Two Functions

$$\frac{d}{dx}[f(x)\,g(x)] = \left[f(x)\frac{d}{dx}g(x) + g(x)\frac{d}{dx}f(x)\right] \quad (1.32)$$

1.5.6 Differential Coefficient of a Quotient of Two Functions

$$\frac{d}{dx}\left[\frac{f(x)}{g(x)}\right] = \frac{\left[g(x)\frac{d}{dx}f(x) - f(x)\frac{d}{dx}g(x)\right]}{[g(x)]^2} \quad (1.33)$$

Problem 10. Find $\frac{d}{dx}x^7$.

Solution: Here, $y = x^7$

$$\Rightarrow \quad \left(\frac{dy}{dx}\right) = \frac{d}{dx}x^7 = 7 \cdot x^{7-1} = 7x^6$$

Problem 11. If $y = 5x^3 + 3\sin x$, find $\left(\frac{dy}{dx}\right)$.

Solution: Here,

$$\left(\frac{dy}{dx}\right) = \frac{d}{dx}(5x)^3 + \frac{d}{dx}(3\sin x)$$

$$= 5 \times 3x^2 + 3(\cos x)$$

$$= 15x^2 + 3\cos x$$

Problem 12. If $y = \sin x^3$, find $\left(\frac{dy}{dx}\right)$.

Solution: Putting $x^3 = t$ *in* equation $y = \sin x^3$, we get $y = \sin t$

Now

$$\frac{dt}{dx} = \frac{d}{dx}x^3$$

$$= 3x^2$$

and $$\left(\frac{dy}{dt}\right) = \frac{d}{dt}\sin t = \cos t$$

Here, $$\frac{dy}{dx} = \frac{dy}{dt} \times \frac{dt}{dx}$$
$$= \cos t \times 3x^2$$
$$= 3x^2 \cos x^3.$$

1.5.7 Partial Differentiation

Suppose $z = f(x, y)$ is a function of two variables x and y. Then, x and y are independent variables and z is called the dependent variable. If differential coefficient, *i.e.*, derivative of a function of several independent variables can be found with respect to any one of them, keeping the other variables as constant, we say it to be **a partial derivative**.

Thus, **partial differentiation** is the operation of finding the partial derivative of a function of more than one independent variables.

Hence, $\frac{\partial z}{\partial x}, \frac{\partial z}{\partial y}$ are the partial differential coefficient of z with respect to x, y, etc.

Second Order Partial Differential Coefficient

Suppose $z = f(x, y)$; then, $\frac{\partial z}{\partial x}$ and $\frac{\partial z}{\partial y}$ are still the functions of x and y, and can be further differentiated partially. Thus, we get **second order partial derivatives** of z like

$$\frac{\partial}{\partial x}\left(\frac{\partial z}{\partial y}\right), \frac{\partial}{\partial y}\left(\frac{\partial z}{\partial x}\right), \frac{\partial}{\partial x}\left(\frac{\partial z}{\partial x}\right), \frac{\partial}{\partial y}\left(\frac{\partial z}{\partial y}\right) \text{ or } \frac{\partial^2 z}{\partial x \partial y}, \frac{\partial^2 z}{\partial y \partial x}, \frac{\partial^2 z}{\partial x^2} \text{ and } \frac{\partial^2 z}{\partial y^2}.$$

We can also establish that $$\frac{\partial^2 z}{\partial x \partial y} = \frac{\partial^2 z}{\partial y \partial x}. \tag{1.34}$$

Problem 13. If $z = f(x, y) = x^4 + 2x^2y^2 + y^4$ find the values of

$$\frac{\partial z}{\partial x}, \frac{\partial z}{\partial y}, \frac{\partial^2 z}{\partial x \partial y}, \frac{\partial^2 z}{\partial y \partial x}, \frac{\partial^2 z}{\partial x^2}, \frac{\partial^2 z}{\partial y^2}$$

Solution: Given that $z = x^4 + 2x^2y^2 + y^4$, we have

$$\frac{\partial z}{\partial x} = 4x^3 + 4xy^2$$

$$\frac{\partial^2 z}{\partial x^2} = 12x^2 + 4y^2$$

$$\frac{\partial^2 z}{\partial x \partial y} = 8xy$$

$$\frac{\partial^2 z}{\partial y \partial x} = 8xy$$

$$\frac{\partial z}{\partial y} = 4x^2y + 4y^3$$

$$\frac{\partial^2 z}{\partial y^2} = 4x^2 + 12y^2.$$

1.6 MAXIMA AND MINIMA

Procedure and working method for obtaining maxima and minima of $f(x)$ is based on the following points:

1. If $y = f(x)$, first we have to get $\left(\frac{dy}{\partial x}\right) = f'(x)$ and then equate it to zero.
2. On solving the above equation for x, we get the roots, say a_1, a_2; thus, $x = a_1$, $x = a_2$ are the points at which $f(x)$ can have maximum or minimum, provided $f(x)$ is stationary at $x = a_1$, $x = a_2$, etc.
3. Next we have to get $\frac{d^2 y}{dx^2} = f''(x)$ and substitute it by turns with $x = a_1$, $x = a_2$, etc.
4. If $d^2y/dx^2 = f''(x) = f''(a_1)$ is negative, we have the maxima at $x = a_1$. On the other hand, if $d^2y/dx^2 = f''(x) = f''(a_1)$ is positive, we have the minima at $x = a_1$.
5. If $(d^2y/dx^2) = f''(x) = f''(a_1) = 0$, find $(d^3y/dx^3) = f'''(x)$ and put $x = a_1$ and now if $f'''(a_1) \neq 0$, there is neither a maximum nor a minimum at $x = a_1$.

We can continue further in the same way and repeat the process on other roots as well.

Problem 14. Find the maximum and minimum values of the function

$$y = x^5 - 5x^4 + 5x^3 - 1$$

Solution: Here, $(dy/dx) = 5x^4 - 20x^3 + 15x^2$

Putting $(dy/dx) = f'(x) = 0$ for maximum and minimum, we have

$$5x^4 - 20x^3 + 15x^2 = 0$$

$$\Rightarrow 5x^2 (x^2 - 4x + 3) = 0$$

$$\Rightarrow 5x^2 (x - 1)(x - 3) = 0$$

The above equation gives $x = 0, 1, 3$ as the possible roots.

Now $d^2y/dx^2 = f''(x) = 20x^3 - 60x^2 + 30x$

At $x = 1$, $\left(\frac{d^2y}{dx^2}\right) = f''(x) = f''(1) = 20(1)^3 - 60(1)^2 + 30(1) = -10$

(a negative value).

Hence, value of the function ($y = x^5 - 5x^4 + 5x^3 - 1$) has maxima at $x = 1$, and the maximum value of the function.

$$y\,(\text{max}) = (1)^5 - 5(1)^4 + 5(1)^3 - 1 = 1 - 5 + 5 - 1 = 0$$

At $x = 3$, $(d^2y/dx^2) = 20(3)^3 - 60(3)^2 + 30(3) = 90$ (a positive value).

Hence, value of the function has minima at $x = 3$, and the minimum value of the function.

$$y\,(\text{min}) = (3)^5 - 5(3)^4 + 5(3)^3 - 1$$
$$= -28.$$

1.7 RECIPROCITY RELATIONS

In certain chemical laws and processes, the quantities are related through inverse dependence. These type of relationships are termed reciprocity relations. Some examples are given below:

1. **Boyle's Law:** We know that at constant temperature, pressure of a given mass of gas is inversely proportional to volume of the gas

   i.e., $$P \propto \frac{1}{V} \text{ (at constant } T)$$

   or $$PV = \text{constant} \tag{1.35}$$

   This statement can be graphically represented as follows Fig. 1.6.

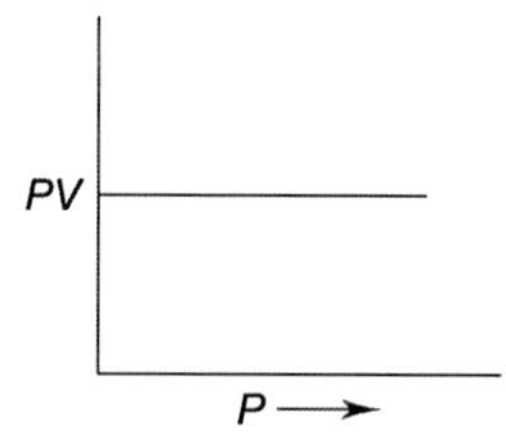

Fig. 1.6 Graphics representation.

2. **Ostwald's Dilution Law:** Degree of dissociation of a weak electrolyte, at constant temperature and for a given solvent, is inversely proportional to the square-root of its concentration.

   i.e, Degree of Dissociation (α) $\propto \frac{1}{\sqrt{c}}$.

1.8 INTEGRATION OF A FUNCTION

If $\frac{d}{dx}F(x) = f(x)$; then, $F(x)$ is called the integral of $f(x)$ with respect to x. By definition,

$$\int f(x)\,dx = F(x) + C \tag{1.36}$$

where, C is an arbitrary constant, independent of x. In Eq. 1.36, the function $f(x)$ is called "integral" and the process is called "integration". This is called indefinite integral of $f(x)$.

Definite Integral: If the range of variable x is specified by the interval (a, b), and $f(x)$ is a single value and continuous in (a, b), then a is called the lower limit of x, and the integral is

$$\int_a^b f(x)\,dx = \left(F(x)\right)_a^b = F(b) - F(a). \tag{1.37}$$

1.8.1 Integration of Some Useful Functions

1. $\int x^n\,dx = \dfrac{x^{n+1}}{n+1}$, when $n \neq (-1)$
2. $\int \dfrac{1}{x}\,dx = \log_e x$
3. $\int \cos x\,dx = \sin x$
4. $\int \sin x\,dx = -\cos x$
5. $\int e^x\,dx = e^x$
6. $\int a^x\,dx = \dfrac{a^x}{\log_e x}$

Problem 15. Evaluate $\int 4 \sin x\,dx$

Solution: $\int 4 \sin x\,dx = -4\cos x + c$.

Problem 16. Evaluate $\int_0^1 (x+1)\,dx$

Solution:

$$\int_0^1 (x+1)\,dx = \int_0^1 x dx + \int_0^1 1 . dx$$

$$= \left[\frac{x^2}{2}\right]_0^1 + [x]_0^1 = \left(\frac{1}{2} - 0\right) + (1-0) = \frac{3}{2}.$$

Problem 17. Evaluate $\int \left(\dfrac{5x^4}{1+x^5}\right) dx$

Solution: Let $(1 + x^5) = t$

Then, $5x^4\,dx = dt$

Hence

$$\int \frac{5x^4}{\left(1+x^5\right)} = \int \frac{dt}{t} = \log_e t + \text{constant } (c)$$

Replacing t by $(1+x^5)$, we get

$$\int \frac{5x^4}{(1+x^5)} = \log_e (1+x^5) + c.$$

1.9 PERMUTATION AND COMBINATION

1.9.1 Factorials

Factorial n is the continued product of first n natural numbers.

It is written as $\underline{|n}$ or $n!$ and is expressed as

$$n! = n \times (n-1) \times (n-2) \times \qquad (1.38)$$

Thus, Factorial 6, *i.e.*, $6! = 6 \times 5 \times 4 \times 3 \times 2 \times 1$

Also, $$0! = 1 \text{ and } n! = n \times (n-1)! \qquad (1.39)$$

1.9.2 Permutation

It is the number of ways in which n distinct things can be arranged taking "r" at a time, r is such that $1 \le r \le n$.

Symbol of permutation is ${}^np_r$

Mathematically, $${}^np_r = \frac{n!}{(n-r)!} \qquad (1.40)$$

$${}^np_n = n!;\ {}^np_r = n \times {}^{(n-1)}p_{(r-1)}$$

Problem 18. Find the value of ${}^{10}p_4$.

Solution:

$${}^{10}p_4 = \frac{10!}{(10-4)!} = \frac{10 \times 9 \times 8 \times 7 \times 6!}{6!}$$
$$= 5040.$$

Problem 19. In how many ways four students can be seated in two empty chairs.

Solution: Here, $n = 4$ and $r = 2$

Thus, four students can be seated in two empty chairs in ${}^4p_2$ ways.

$$\Rightarrow \qquad {}^4p_2 = \frac{4!}{(4-2)!} = \frac{4 \times 3 \times 2!}{2!} = 12$$

1.9.3 Combination

All the combinations, which can be made of n objects, taking r at a time are denoted by the term ${}^nC_r$, when $0 \le r \le n$.

Mathematically,

$$^{n}C_{r} = \frac{n!}{r!(n-r)!} \tag{1.41}$$

Problem 20. Evaluate $^{5}C_{2} = \dfrac{5!}{2!(5-2)!}$

Solution:

$$^{5}C_{2} = \frac{5 \times 4 \times 3!}{2 \times 1 \times 3!} = 10$$

Problem 21. How many triangles can be formed by joining the vertices of an octagon?

Solution: There are 8 vertices of an octagon. A triangle is formed by joining three vertices, and the selection of groups of 3 vertices from the available 8 vertices can be done in $^{8}C_{3}$ ways.

Hence, the number of triangles = $^{8}C_{3}$

$$= \frac{8 \times 7 \times 6 \times 5!}{3! \times (8-3)!} = \frac{8 \times 7 \times 6}{3 \times 2} = 56.$$

1.10 PROBABILITY

Classically, probability was defined as the ratio of outcomes favourable to a particular event to the total number of equally likely outcomes. For example, the probability of getting an odd number (1, 3, 5) in the throwing of a dice (it is a homogeneous cube with six faces marked with numbers or dots, from 1 to 6, one number or dot on each face) is $\frac{3}{6}$, *i.e.*, $\frac{1}{2}$.

1.10.1 Random Experiments and Sample Space

An experiment is called a random experiment, if all of its outcomes are known in advance, except that which outcome will occur in a particular trial of the experiment cannot be predicted with certainty.

The set S of all possible outcomes is given by $S = \{1, 2, 3, 4, 5, 6\}$ and is called the **sample space** (of a given activity, say throwing a dice). Elements or members of a sample space are called the **sample points**. The activity (such as throwing a dice) is a random experiment.

1.11 EVENT

An event is a sub set of the sample space. Consider, for example, tossing of a coin, we have

$$\text{Sample space } (S) = \{H, T\}$$

where, H, stands for head and T for tail.

In all there are 4 subsets, namely, $\{H\}$, $\{T\}$, $\{HT\}$ and ϕ (the empty subset) in this case each subset of S is an event, which may be

1. **Simple Event:** The event which has only one sample point of the sample space. In the above case, events $\{H\}$ and $\{T\}$ are simple events.
2. **Compound Event:** The event which has more than one sample point. Here, $\{HT\}$ is a compound event.
3. **Sure Event:** For throwing a dice, $S = \{1, 2, 3, 4, 5, 6\}$. Here, set S itself is also a subset of S and, therefore, represents an event that occurs whenever we carry out the activity. Thus, event S is named a sure event.
4. **Impossible Event:** The empty set ϕ is also a sub set of the sample space S; therefore, it also represents an event. Let E be the event, "the number turns up is a cube other than 1". Then, $E = \phi$, and we call the event ϕ as an impossible event. Thus, using a set notation, we can write the probability of occurrence of the event 4 as

$$P(A) = \frac{\text{Number of outcomes favourable to 4}}{\text{Total number of possible outcomes}}$$

$$= \frac{m}{n} \tag{1.42}$$

Using standard notations, we can write

$$P(A) = \frac{n(A)}{n(S)} \tag{1.43}$$

where, $n(S)$ denotes the number of elements (sample points) in the sample space S and $n(A)$ is the number of elements in A.

Problem 22. A coin is tossed twice, what is the probability of occurrence of at least one tail?

Solution: Here, the sample space may be given as

$$S = \{HH, TH, HT, TT\}.$$

For the event E " the occurrence of at least one tail", we have

$$E = \{TH, HT, TT\}$$

and $$P(E) = \frac{\text{Number of outcomes favourable to } E}{\text{Total possible outcomes}} = \frac{3}{4}.$$

Problem 23. There are four men and six women on the district bar council. If one council member is selected for a committee at random, how likely is that a women is selected?

Solution: The total number of persons $= 4 + 6 = 10$. Out of these one person can be selected in ${}^{10}C_1$ ways. Out of six women, one can be selected in ${}^{6}C_1$ ways. Hence, P (one women) $= \frac{{}^{6}C_1}{{}^{10}C_1} = \frac{6}{10} = \frac{3}{5}$.

Objective Type Questions

1. The value of x in $\log x = \bar{1}.3919$ is

(*a*) 0.25 (*b*) 25

(*c*) 0.025 (*d*) 250

2. The equations represented as

$$x = at^2;\ y = 2at \text{ on tracing will give}$$

(*a*) An ellipse (*b*) A parabola

(*c*) A circle (*d*) A straight line

3. The slope of the line $y = 0.7x + 11.285$ is

(*a*) 0.113 (*b*) 0.700

(*c*) 1.0 (*d*) 0.285

4. Differential coefficient of $x^5 \sin x$ with respect to x is

(*a*) $x^5 \cos x + 5\, x^4 \sin x$ (*b*) $x^5 \tan x$

(*c*) $5x^4 \cos x + 3x^2 \sin x$ (*d*) $x^6 \sin x + 5x^4 \cos x$

5. Differential coefficient of log (sin x) is

(*a*) $\cos x$ (*b*) $\cot x$

(*c*) $\tan x$ (*d*) $-\cos x$

6. If $\dfrac{1}{u} = \sqrt{x^2 + y^2 + z^2}$, the expression $\dfrac{d^2u}{dx^2} + \dfrac{d^2u}{dy^2} + \dfrac{d^2u}{dz^2}$ is equal to

(*a*) 1 (*b*) 0

(*c*) – 1 (*d*) 2

7. The function $x^5 - 5x^4 + 5x^3 - 1$ is maximum at

(*a*) $x = 0$ (*b*) $x = -1$

(*c*) $x = +1$ (*d*) $x = 2$

8. Integration of $\int_0^1 (x+1)\, dx$ is equal to

(*a*) $\dfrac{1}{2}$ (*b*) 1

(*c*) $\dfrac{3}{2}$ (*d*) 2

9. The number of triangles which can be formed by joining the vertices of a hexagon are

(*a*) 10 (*b*) 20

(*c*) 30 (*d*) 40

10. The number of different permutations of the word BANANA is

(*a*) 720 (*b*) 60

(*c*) 120 (*d*) 360

ANSWERS

1. (*a*)	**2.** (*b*)	**3.** (*b*)	**4.** (*a*)	**5.** (*b*)
6. (*b*)	**7.** (*c*)	**8.** (*c*)	**9.** (*b*)	**10.** (*b*)

Short Answer Type Questions

1. Use logarithm table and evaluate $\dfrac{\log_{10} 81}{\log_{10} 27}$.

[**Ans.** $\dfrac{4}{3}$]

2. Use logarithm table to evaluate $4\sqrt{0.0234}$.

[**Ans.** 0.3924]

3. Differentiate $y = \sqrt{x^3}$ with respect to x.

[**Ans.** $\left(\dfrac{-3}{2}\right) x^{-5/2}$]

4. If $u = x \log xy$, where $x^3 + y^3 + 3xy = 1$ find $\dfrac{du}{dx}$.

5. Find the maximum value of xe^{-x}.

[**Ans.** $(1/e)$].

6. Find the number of ways in which a committee of 5 members can be selected from 6 men and 5 women.

[**Ans.** 200]

7. Prove that if $x^3 y^3 + 3xy \sin y = e^y$, then $\dfrac{dy}{dx}$ is

$$\frac{\left(3x^2 y^3 + 3 \sin y\right)}{3x^3 y^3 + 3x \cos y - e^y}$$

8. Find the value of integral $\int_0^{\frac{x}{2}} x \cos x \, dx$.

[**Ans.** $\left(\dfrac{\pi}{2} - 1\right)$]

9. If $z = x \log y$, then find the value of $\dfrac{\partial^2 z}{\partial x \, \partial y}$.

[**Ans.** $\left(\dfrac{1}{y}\right)$]

10. If ${}^{10}C_r = {}^{10}C_{r+6}$, find r.

[**Ans.** 2]

Subjective Questions

1. Find the differential coefficient of $3\sqrt{x} + 5\log_e x - 11\log_a x$ with respect to x.

[**Ans.** $\frac{3}{2}x^{-1/2} + \frac{5}{x} \cdot - \frac{19}{x}\log_a e.$]

2. The sum of one number and three times of a second number is 60. Among the possible number satisfy the stated condition, and find the pair whose product is maximum.

[**Ans.** (30, 10)]

3. Evaluate $\int x^n \log x \, dx$

[**Ans.** $\frac{x^{n+1}}{(n+1)}\log x - \frac{1}{(n+1)^2}x^{n+1} + C$]

4. From a class of 20 students, 4 are to be chosen for a competition. In how many ways this can be made?

[**Ans.** 4845]

5. Explain reciprocity relation.

6. Find the diferential coefficient of $\frac{\cos x}{\log_a x}$.

[**Ans.** $-\sin x \log_a x - \frac{1}{x}\cos x \log_a e$].

7. Prove that a conical tent of a given capacity will require the least amount of canvas when the height is $\sqrt{2}$ times the radius of the base.

8. Evaluate $\int e^{ax+b}\, dx$.

[**Ans.** $\frac{1}{a}e^{ax+b} + C$]

9. Solve $\log_{10}(x+1) + \log_{10}(x-1) = \log_{10} 48$.

[**Ans.** $x = \pm 7$]

10. Trace the curve defined by the equation $y^2(a^2 + x^2) = x^2(a^2 - x^2)$.

2

Computers for Chemists

Information technology has brought new revolution in every sphere of life, including chemical science (its practices as well as teaching methodologies). Computer, which forms its most important part, can be said as the "machine of 21st century".

2.1 INTRODUCTION TO COMPUTERS

Computer is a fast electronic data manipulating system designed to accept raw data, store and process them under the instructions of a set of program to produce quite meaningful results. It is able to help us in solving many problems of our life. The speed at which a computer works is very fast and the machine is able to process millions of instructions per second (MIPS) giving very accurate results. A computer can store a large amount of data as well as instructions. The information stored in it can be recalled whenever required.

Modern computers are useful in the field of science and technology, office automation, process control, publishing, manufacturing, medical diagnosis (*e.g.*, CT SCAN), railways, airline traffic control, working of nuclear reactors, computer aided designing (CAD) and computer aided manufacturing (CAM), *etc.*

2.2 TYPES OF COMPUTERS

The computer industry classifies computers as micro (PCs), mini, mainframe and super computers.

2.2.1 Minicomputers

Most of the minicomputers introduced in 1970s operated on 16-bits. These are able to support a large variety of fast operating peripheral units. More than hundred terminals can be served by them and a better communication link can be established between workstations and a central miniframe computer. Mini operating system program provides a better environment than that of PCs.

2.2.2 Mainframe Computers

These computers can handle a data length of more than 64 bits and have quite a big library application program and a very high level programming language.

2.2.3 Super Computers

These are the most powerful computers which use gallium arsenide as the chip material. The processors used in them are (*a*) pipeline processors, *e.g.,* in CRAY 1a, VP200 and (*b*) multiprocessors, *e.g.,* in CRAY XMP and CRAY 2. Highly complex problems, *e.g.,* weather forecasting, computer aided designing and nano-technology experiments are handled by super computers. Here, each address location holds 64 bits of information and the cycle-time required is as low as a few nanoseconds.

2.2.4 Micro Computers – Personal Computers — IBM PC

Z × 80 was the first home computer developed in the year 1982. ROM (Read Only Memory) had a memory of 16 KB in an interpreter meant for BASIC with RAM (Random Access Memory) of 128 KB capacity. Computer had connectivity to CTV, printer, keyboard and power supply.

First personal computer (PC) with 16 bits was made available in seventies. IBM PC was introduced around 1981 and had a memory of 256 KB; it accommodated two floppy drives (with storage capacity of 160 KB).

Today's PC with peripherals and CPU, *e.g.,* operating system, hard-disk, CD device, keyboard, printer, monitor, mouse, *etc.*, is shown below in Fig. 2.1.

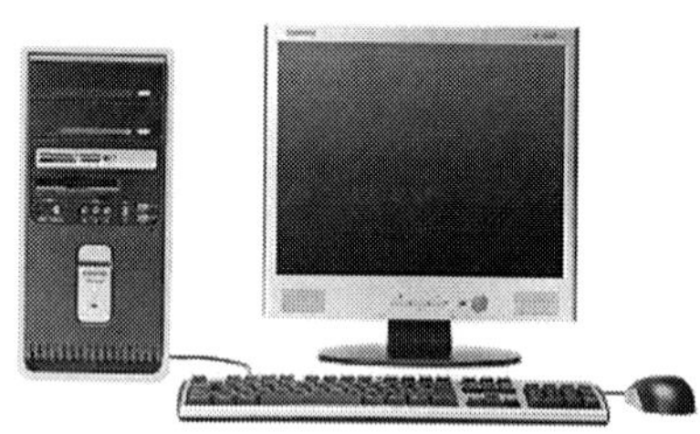

Fig. 2.1 Personal Computer PC.

2.3 COMPONENTS OF A COMPUTER

Computer is a system comprising two distinct parts. These are:

1. **Hardware:** It consists of several electrical and mechanical devices such as keyboard, floppy-disk and compact-disk drive (E), monitor, hard-disk, printer, mouse *etc.*
2. **Software:** A set of instructions as fed to the computer memory so that it is able to perform and provide the desired results.

2.4 COMPUTER HARDWARE–INPUT DEVICES

1. **Keyboard:** A PC keyboard is shown in Fig. 2.2. Keys on the board are arranged almost similar to the keys of a typewriter with *A* to *Z* English alphabets, special function keys, numbers and some special functions, *e.g.,* +, -,? along with Page Dn, Page Up-arrow keys, enter-key, *etc.*

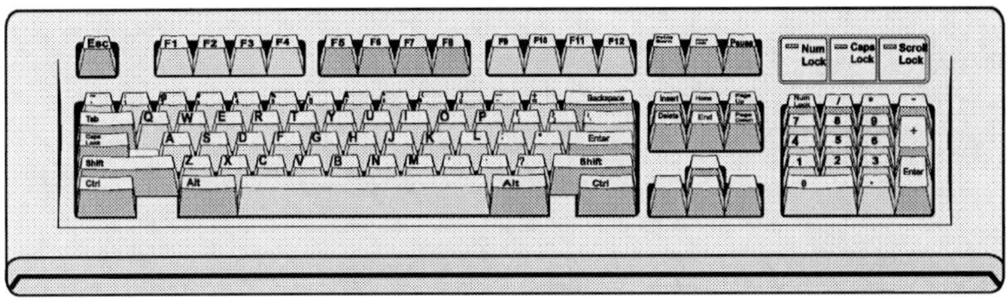

Fig. 2.2 Keyboard of a PC.

2. **Floppy (A) and Compact-Disk drives (E) :** A circular floppy (5.25 in.) or CD drives are also provided so as to work with floppy-disk (1.44 MB) or CD (700 MB) (Recordable)/data/audio/video storage systems.
3. **Magnetic Ink Character Reader (CARD) :** This is used in banks for delivering M.I.C.R. cheques, in which magnetic ink is used to print numbers, *etc.*, on them.

2.5 COMPUTER HARDWARE–OUTPUT DEVICES

1. **Visual Display Unit (V.D.U.)–Monitor-Screen:** Just like television, computers also have a cathode ray tube (CRT) display monitor, which flashes the output data/picture on the screen.
2. **Printers:** Permanent (hard) copy of the output data from a computer can be printed by using a DOT MATRIX/INK-JET/LASER/LINE printer. High quality letters, numbers and graphics can be created for use depending on the software being used.

2.6 COMPUTER HARDWARE–STORAGE DEVICES: HARD DISK

This is a device, designated as © drive, made available with a storage capacity 40 GB or more. We can read/write and obtain information entered in any part of the disk randomly.

2.7 BINARY NUMBERS AND COMPUTER ARITHMETIC

There are two kinds of data which is available with us

1. **Character Data:** Where we present alphabets, special symbols such as +, –, *, *etc.*
2. **Numeric Data:** Where we use digits from 0 to 9.

The data is fed to a computer in the external format, which then processes it into internal format (0 and 1) to give us the desired results. It is, therefore, necessary to establish a certain relationship between the external and internal formats of data accessed by a computer.

2.7.1 Binary Number System

Computer system is basically binary, *i.e.*, depend on two digits (0 and 1) only.

Decimal Number	*Binary Number*	*Conversion for Binary*
0	0	0
1	1	$1 \times 2^0 = 1$
2	10	$1 \times 2^1 + 0 \times 2^0 = 2$
3	11	$1 \times 2^1 + 1 \times 2^0 = 3$
4	100	$1 \times 2^2 + 0 \times 2^1 + 0 \times 2^0 = 4$
5	101	$1 \times 2^2 + 0 \times 2^1 + 1 \times 2^0 = 5$
6	110	$1 \times 2^2 + 1 \times 2^1 + 0 \times 2^0 = 6$
7	111	$1 \times 2^2 + 1 \times 2^1 + 1 \times 2^0 = 7$
8	1000	$1 \times 2^3 + 0 \times 2^2 + 0 \times 2^1 + 0 \times 2^0 = 8$

In order to convert any decimal number into a binary number, it is divided by 2, which gives us a remainder of either '0' or '1'. We find it to be the least significant figure of binary number. The quotient is again divided by 2 and the remainder now obtained is made the second digit of binary number, and so on.

■ **Example:** Conversion of decimal number 25 into binary number.

First proceed as follows:

Number	*Quotient*	*Remainder*
25/2	= 12	1 (the least significant digit)
12/2	= 6	0 (second digit of the binary number)
6/2	= 3	0 (third digit of the binary number)
3/2	= 1	1 (fourth digit of the binary number)
1/2	= 0	1 (the most significant digit)

Now place all the remainders in a sequence, *i.e.*, starting from the most significant to the least significant digit to get the binary number. Thus,

Decimal number $(25)_{10}$ = Binary number $(11001)_2$

Binary number is written in bracket indicating the base outside the bracket.

To convert a decimal number having decimal point in between, the number after the decimal point is multiplied by 2, and, the digit before the decimal in the product is the most significant digit of the binary number. Now again the number after the decimal is multiplied by 2 and the number after the decimal is the second most significant digit.

This procedure is continued until the value of the product is such that the part after the decimal become zero.

■ **Example:** Conversion of a decimal number 0.53125 into binary number.

First proceed as follows:

Number	*Product*	*Digit Before Decimal*
0.53125×2	= 1.0625	1 (least sgnificant digit)
0.0625×2	= 0.1250	0
0.1250×2	= 0.2500	0
0.250×2	= 0.500	0
0.500×2	= 1.000	1 (most significant digit)

Now place the digits before the decimal, starting from the most significant digit to the least significant digit. Thus,

Decimal number $(0.53125)_{10}$ = Binary number $(10001)_2$

A binary number can be reconverted into the decimal number. It is understood by the following example:

■ **Example:** Conversion of binary number $(11001)_2$ into the corresponding decimal number.

First proceed as follows:

Place	*Digit*	*Value of the Binary Position*	*Value of the Digit*	*Value in the Decimal System*
The least significant digit	1	$2^0 = 1$	1×1	1
	0	$2^1 = 2$	0×2	0
	0	$2^2 = 4$	0×4	0
	1	$2^3 = 8$	1×8	8
	1	$2^4 = 16$	1×16	16
			Total value	25

Thus, Binary Number $(11001)_2$ = Decimal Number $(25)_{10}$.

2.8 INTRODUCTION TO COMPUTER LANGUAGE

Every computer has a control program called Operating System, which takes care of operations of input, output and other secondary devices. The "operating system program" used in earlier version of PCs was provided through a floppy disk, and hence was denoted as Disk Operating System (DOS). Later on, when Intel 8080 microprocessor was used, MS DOS (Microsoft Corporation Disk Operating System) was introduced. MS DOS, in the latest versions of Windows XP Professional, is provided in hard-disk drive and starts functioning as soon as the power to the computer is switched on taking only a few seconds, Command Prompt C:\>- appears on the screen, and DOS Command is loaded into the main memory. We can now open, any **Application Program,** present as **icon**, on the screen, by clicking through **mouse or keyboard using the Menu Bar**. The most popular and latest versions of operating system packages are MS-Windows-XP Professional, Linux and Unix.

2.8.1 Computer Languages

The set of instructions, provided to a computer so as to get the desired task completed, is called a Program. Computer understands only **Machine Language**, *i.e.*, the set of instructions written using binary numbers, meant for the presence or absence of electrical pulse. A program written in machine language is called Object Program and these are quite tedious and time consuming. Later on, symbolic words, *e.g.*, MOV standing for move, were used and their machine language translation was done automatically by operating system using Assembly Language. However, both of them were low level languages used in computer. A program translating the actual words into binary pattern, called Assembler, was later developed.

2.8.2 High Level Languages

In these languages, a program written in English, called source program, *e.g.*, formula

$$\text{Force } (F) = \text{Mass } (m) \times \text{Acceleration } (a)$$

was converted into binary pattern – called Object Program using translator (Compiler or Interpreter). In high level languages, such as FORTRAN (meant for scientific calculations), Pascal (for general and educational purposes), C (system programming using Unix operating system), an interpreter translates the program. Compiler translates program into object program as a whole and is done quite fast. Any error done is corrected by Editor before the program is finally executed.

2.9 PROGRAMMING

In order to solve a problem, one should be able to completely understand the problem, know the best pathway to solve it with requisite data to start and also the nature of final results. The solution of a problem involves the following steps:

2.9.1 Develop an Algorithm

Here one should divide the whole problem into logical steps which are to be followed to arrive at the solution, *e.g.*, suppose we have to apply the Phase Rule

$$F = C - P + 2$$

To calculate the Degrees of Freedom (F) of a particular system:

Step 1: We should know the number of components (C) and number of phases (P) present in the system.

Step 2: We have to develop grammar (language) to attain ($C - P$), for different values of C and P.

Step 3: To add a number (2) to the result(s) thus obtained.

2.9.2 Flow Chart

Diagrammatic presentation of the algorithm is known as flow chart.

2.9.3 Programming in Basic Language

Basic is the acronym for "Beginners All-Purpose Symbolic Instruction Code".

In basic program every **statement** has a line number and two statements cannot have the same line number.

Each statement includes alphabets and some special characters, such as:

Only capital English letters, *e.g.*, A, B, C, ..., Z are used, decimal digits 0 ... 9 are used and special characters +, –, (;), #, $, %, =, / are used for addition, subtraction, semicolon, numeric, string variables, percentage, assignment of relation of equality, division operator, respectively.

Basic program consists of different types of statement, such as:

1. INPUT Statement: provide data to computer : READ–to allow computer to store the data in memory.
2. REM (REMARK) Statement: to introduce comments.
3. LET Statement: to assign string or numerical value to a string or numerical variable.
4. PRINT statement: to transfer string or numerical output from computer.
5. STOP statement: to terminate the computation.
6. END statement: to indicate the end of a basic program.

Problem 1. Write the BASIC PROGRAM for calculating Degrees of Freedom from the Phase Rule.

Solution: 10 REM CALCULATION OF DEGREE OF FREEDOM (F) OF A SYSTEM USING PHASE RULE

20 REM $F = C - P + 2$

F is not defined

30 READ C, P

40 $F = C - P + 2$

60 DATA 2, 3

80 PRINT "THE DEGREE OF FREEDOM F IS =" ; F

100 STOP

120 END

Problem 2. Write the BASIC PROGRAM for calculation of radius of rth Bohr orbit using the formula (r) = $\frac{(n\lambda)}{(2\pi)}$.

Note : The basic variables for n, λ and π are R%, WAVELENGTH and PI and r is represented by R!

Solution: 10 REM RADIUS OF BOHR ORBIT USING WAVELENGTH FOR ELECTRON FROM DE BROGLIE EQUATION

20 READ WAVELENGTH, *N%*

30 PI = $\frac{22}{7}$

40 *R*! = *N%* * WAVELENGTH / (2 * *PI*)

80 PRINT "THE RADIUS OF THE" = ; R!

100 PRINT "ORBIT NUMBER" ; *N%*

```
120 DATA   3.325, 1
130 STOP
140 END
    OUT-PUT
    THE RADIUS OF THE BOHR ORBIT = 5289 E–04
    ORBIT NUMBER = 1
```

Note : The value calculated by computer program, *i.e.*, 5289 E–04 corresponds to 0.5289 Å.

2.10 LAPTOP COMPUTER

A laptop computer, usually called a notebook computer by manufacturers, is a battery– or AC-powered personal computer. These are smaller than even a briefcase, and can easily be transported and conveniently used in temporary spaces such as on airplanes, in libraries, in temporary offices, and at meetings. A laptop typically weighs less than 5 pounds and is 3 in. or less in thickness. Among the known makers of laptop computers are IBM, Apple, Compaq, Dell, and Toshiba. Most of them have following features:

Processor — Intel
Processor Speed (MHz) — 2000

Standard RAM (MB) — 2048
Inbuilt Hard Disk Drive Capacity (GB) — 160
Display Screen Size (in.) — 15.4
Internal Modem
Modem Speed (Kbps) — 56
Microphone
Speakers
Operating System — DOS / WINDOWS
Type of Mouse — Touchpad
Width (mm) — 360
Depth (mm) — 263
Height (mm) — 34.9
Weight (kg) — 2.8
Battery
Battery (type) — Li Ion
Battery Life (hours) — 2.5

Laptop computers generally cost more than the desktop computers with the same capabilities because these are most difficult to design and manufacture. A laptop can effectively be turned into a desktop computer with a docking station, a hardware frame that supplies connections for peripheral input/output devices such as a printer or larger monitor. Laptops usually come with displays that use thin-screen technology. The thin film transistor or active matrix screen is brighter and views better at different angles than the STN or dual-scan screen. Laptops use several different approaches for integrating a mouse into the keyboard, including the touch pad, the trackball and the pointing stick. A serial port also allows a regular mouse to be attached. The PC Card is an insertable hardware for adding a modem or network interface card to a laptop. CD-ROM and digital versatile disk drives may be built-in or attachable.

2.11 INTERNET

The Internet Society is an independent international non-profit organisation founded in 1992 to provide leadership in Internet-related standards, education and policy around the world.

The Internet is a global system of networking through which every one, who has connected his/her computer through a standardized internet protocol (IP), could communicate with others. Millions of people, belonging to private business or governmental organizations are interconnected through this network. Internet connections are provided through various cables, fibre optic cables and with so many other technologies involving even wireless connections through ERNET, *etc.*

The information available through internet include e-mail service, file transfer and file sharing, online games and so many other scientific and technological resources of the world wide web (www).

The global data communication is using so many softwares and hardwares while establishing connectivity between a service provider and a user.

We can use internet if we have: (*i*) a Pentium computer with at least 32 MB RAM, (*ii*) a CD-ROM drive, (*iii*) a MODEM with minimum 33 kbps speed, (*iv*) Windows 98 or Windows XP operating system, (*v*) a telephone link and (*vi*) an IP account, through BSNL, *etc.*

IP account holder is given a password, through which, he/she can access the internet on the computer using the telephone connectivity.

Web Address: A web address starts with http://name of the site.subdomain.domain.

Example : (http://www.sancharnet.in) is a website of Bharat Sanchar Net.

(www.ignou.ac.in) is the website of Indira Gandhi National Open University.

Objective Type Questions

1. A computer does not have:

(*a*) Input unit (*b*) Central processing unit

(*c*) Person to read English (*d*) Output unit

2. Digital computer uses binary digits, which are
   (*a*) 1, 2 (*b*) 0, 1
   (*c*) 3, 4 (*d*) 1, 1
3. Capacity of memory is measured in terms of bytes, kilo bytes, megabytes or gigabytes. One giga byte is:
   (*a*) 1024 bytes (*b*) 1024×1024 bytes
   (*c*) $1024 \times 1024 \times 1024$ bytes (*d*) $(1024)^5$ bytes
4. Which of the following cannot be termed as an operating system ?
   (*a*) DOS (*b*) UNIX
   (*c*) XENIX (*d*) WINDOWS
5. As soon as we press a key on the keyboard, a character appears on
   (*a*) ALU (*b*) CPU
   (*c*) VDU (*d*) OU
6. Which of the following languages computer can directly understand?
   (*a*) MACHINE CODES (*b*) BASIC
   (*c*) FORTRAN (*d*) COBAL
7. A Laser printer is:
   (*a*) an impact printer (*b*) non-impact printer
   (*c*) slow printer (*d*) none of these
8. An object program is a translation of a .... into machine language
   (*a*) Assembly language (*b*) Source program
   (*c*) Backup of program (*d*) High-level language
9. Which of the following is essential for using computer effectively?
   (*a*) Executive software (*b*) Command software
   (*c*) System software (*d*) Utility software
10. High-level language is not translated into machine language by a
    (*a*) Assembler (*b*) Compiler
    (*c*) Interpretor (*d*) Virus

ANSWERS

1. (*c*)	**2.** (*b*)	**3.** (*c*)	**4.** (*c*)	**5.** (*c*)
6. (*a*)	**7.** (*b*)	**8.** (*b*)	**9.** (*c*)	**10.** (*d*)

Short Answer Type Questions

1. What is the difference between a floppy disk and a hard disk?
2. Write the functions of keyboard.

3. What is internet?
4. What is an operating system?
5. What is a language processor?
6. What are application programs?
7. What are utility programs?
8. What is a high-level language?
9. What is a CD-ROM?
10. What is an algorithm?

Subjective Questions

1. Convert a decimal number 0.5625 into the binary number.

[**Ans.** $(1009)_2$]

2. Convert a binary number $(101011)_2$ into the decimal number.

[**Ans.** $(43)_{10}$]

3. Draw a flow chart showing addition of two numbers.

[**Ans.** The flow chart given below]

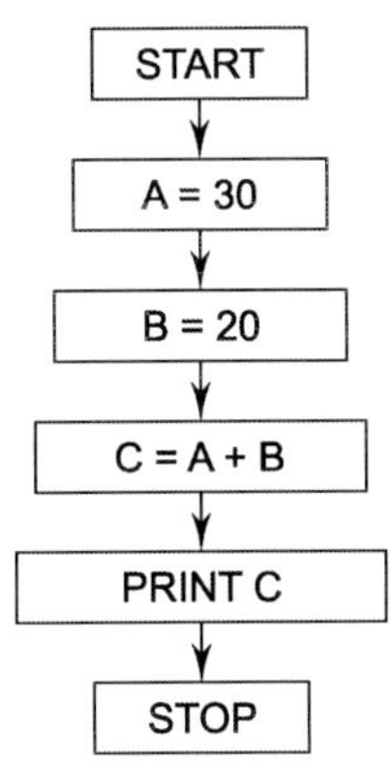

4. Give a program to convert °C to °F using BASIC language.

[**Ans.** PROGRAM IN BASIC written below]

10 REM TO CONVERT CENTIGRADE TO FAHRENHEIT

20 READ CEN

30 FAR = $\left(\frac{9}{5}\right)$*. CEN + 32

40 PRINT FAR = FAR not defined

50 DATA 80

60 STOP

70 END

(For 80°C temperature).

5. Write a program for calculating molecular weight of CHOLESTEROL (using Basic Language).

[**Ans.** PROGRAM IN BASIC is written below]

```
10 REM    MOLECULAR WEIGHT OF CHOLESTEROL
20 INPUT "Name of the organic compound 1" ; Name, COMP. 1 $
30 INPUT  "Give the atomic weights of C, H, O, N", A. WT. CAR, A. WT.
   HYD, A . WT. OXY, A. WT. NIT
40 PRINT
50 MOL. WT. 1 = 27 * A. WT. CAR + 46 * A. WT. HYD + A. WT. OXY
60 PRINT "Name of Compound", "Molecular Weight"
70 PRINT
80 PRINT NAME. COMP. 1 $. MOL. WT. 1
90 STOP
100 END
```

❑❑❑

UNIT-II

1

Gaseous State

1.1 INTRODUCTION

Matter exists in three physical states, *viz.*, the gas, the liquid and the solid. Gases fill up the whole vessel in which they are placed and exert pressure on the walls of the vessel. They can be compressed by applying the pressure and can be liquefied by cooling along with compression. The well-known laws of gaseous behaviour are Boyle's law, Charle's law, Graham's law, Avogadro's law and Dalton's law. There was no theoretical background to justify these empirical generalizations while they were enunciated through simple experimental observations. However, the ideas developed by Joule (1848), Knonig (1856), Clausius (1857), Maxwell and Boltzmann (1859) laid down the basis of Kinetic Molecular Theory of Gases and this gave conclusive explanation to the various gas laws.

1.1.1 Postulates of the Kinetic Theory of Gases

The various postulates of the kinetic theory of gases are:

1. Gases are composed of a large number of minute particles called molecules or atoms (in the case of noble gases).
2. Molecules are quite small and their actual volume is negligible compared to the total volume (space) occupied by the gas.
3. Molecules of the gas in a container are continuously, rapidly and randomly moving in every possible direction and, during their motion, they collide with each other as well as with the walls of the container in which they are filled.
4. The collisions of these molecules with themselves or with the walls of the vessel are perfectly elastic so that there is no loss of energy during the gaseous molecular collisions or impact on the walls of the container.
5. The intermolecular distances in the gaseous state are so large that the

molecules do not exert any appreciable force of attraction or repulsion between themselves and thus move completely independent of another.

6. The pressure exerted by the gas molecules on the walls of the container is due to their bombardment on them.
7. The average kinetic energy of the molecules of the gas is directly proportional to the absolute temperature of the gas.

All these postulates are valid only for ideal gases. However, the real gases approximately obey these generalizations.

1.1.2 Pressure of an Ideal Gas

Consider a small volume of a gas contained in a cubical vessel of edge length *l* cm and, say, there are *n* number of molecules of the gas, each having a mass of *m* Fig. 1.1.

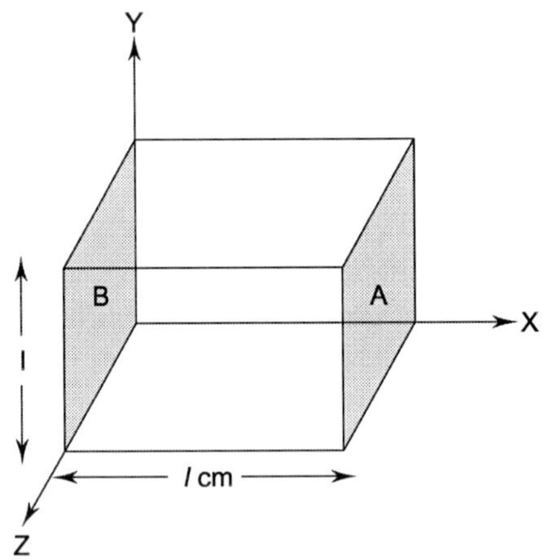

Fig. 1.1(*a*) Cubical vessel containing molecules of gas.

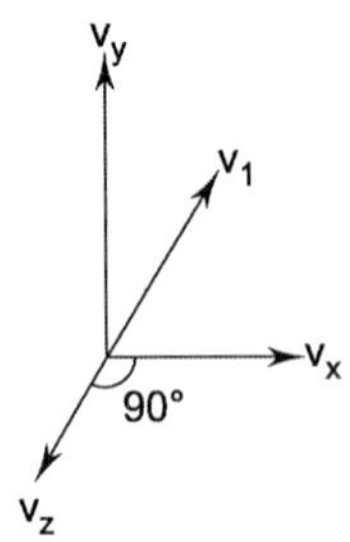

Fig. 1.1(*b*) Resolution of molecular velocity.

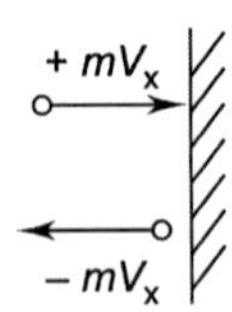

Fig. 1.1(*c*) Transfer of momentum when molecule collide with wall.

Now, according to kinetic theory of gases, the molecules of the gas are moving in all the directions with different velocities. Consider one molecule of the gas having a velocity v_1 cm sec^{-1} in any direction. The velocity v_1 cm sec^{-1} can be resolved into three mutually perpendicular components v_x, v_y and v_z along x-axis, y-axis and z-axis, respectively, as shown in Fig. 1.1 (b). Then,

$$v_1^2 = v_x^2 + v_y^2 + v_z^2 \tag{1.1}$$

Consider the motion of one molecule along the x-axis, striking the wall (A) which is perpendicular to its motion. Since the collision is elastic and the wall remains stationary, only the sign of the velocity component changes. The resulting change in momentum of the molecule in the x-direction Fig. 1.1 (c) in one collision is given as

$$(mv_x) - (-mv_x) = 2mv_x \tag{1.2}$$

Immediately after this collision, the molecule moves in opposite direction to the distance l cm and collide with the opposite wall after $\frac{l}{v_x}$ sec. Thus, it takes $\left(\frac{2l}{v_x}\right)$ seconds to strike against the same wall. So, the number of collisions which take place on this wall in 1 sec due to the velocity component v_x

$$= \frac{v_x}{2l} \tag{1.3}$$

As the change in momentum in one collision is $2mv_x$, the change of momentum per molecule per second due to component v_x for collisions on this wall (*i.e.,* A)

$$= 2mv_x \times \frac{v_x}{2l} = \frac{mv_x^2}{l} \tag{1.4}$$

Similarly, the change in momentum per molecule per second due to component v_x on the wall (B)

$$= \frac{mv_x^2}{l} \tag{1.5}$$

Therefore, the total change in momentum per second per molecule in the x-direction

$$= \frac{2mv_x^2}{l} \tag{1.6}$$

In the same way, changes in the momentum per molecule per second due to impact of gas molecules on the walls in y-direction and z-direction would be $\frac{2mv_y^2}{l}$ and $\frac{2mv_z^2}{l}$, respectively.

Thus, the total change in momentum per second per molecule due to impacts on all the six walls of the cube would be

$$= \frac{2mv_x^2}{l} + \frac{2mv_y^2}{l} + \frac{2mv_z^2}{l} \tag{1.7}$$

$$= \frac{2m}{l}\left(v_x^2 + v_y^2 + v_z^2\right) \quad (1.8)$$

$$= \frac{2m}{l} v_1^2 \qquad \text{[From Eq. 1.1]} \quad (1.9)$$

Here, it is assumed that the number of gaseous molecules possessing velocity v_1 is n_1, velocity v_2 is n_2 and so on. Thus, the change in momentum in 1 sec due to collisions of all the molecules on all the walls would be

$$= \frac{2mn_1 v_1^2}{l} + \frac{2mn_2 v_2^2}{l} + \frac{2mn_3 v_3^2}{l} + \ldots \quad (1.10)$$

$$= \frac{2m}{l} \times n\left(\frac{n_1 v_1^2 + n_2 v_2^2 + n_3 v_3^2 + \ldots}{n}\right) \quad (1.11)$$

where $n = (n_1 + n_2 + n_3 + \ )$, the total number of gas molecules in the cubical container.

The change in momentum [Eq. 1.11] in a simpler form may be written as

$$= \frac{2m}{l} \times nC^2 \quad (1.12)$$

where, $C = \sqrt{\dfrac{n_1 v_1^2 + n_2 v_2^2 + \cdots}{n}}$ termed as the root-mean-square velocity.

As the change in momentum per second equals the force as per Newton's second law of motion, the average force F exerted by the gas molecules may be given as

$$F = \frac{2mnC^2}{l} \quad (1.13)$$

Further, the force is related to pressure as

$$\text{Pressure} = \left(\frac{\text{Force}}{\text{Area}}\right)$$

where area of all the faces of the cubical vessel equals $6l^2$.

The pressure (P) exerted on the walls of the cubical vessel will be given as

$$P = \frac{2mnC^2}{l} \times \frac{1}{6l^2} = \frac{1}{3}\frac{mnC^2}{l^3} \quad (1.14)$$

Here, $l^3 = V$, the volume of the cubical container (or the gas).

So, $$P = \frac{1}{3}\frac{mnC^2}{V}$$

$$\Rightarrow \qquad PV = \frac{1}{3} mnC^2 \quad (1.15)$$

This is called **kinetic gas equation.**

Problem 1. Derive (*i*) Boyle's law, (*ii*) Charle's law, (*iii*) Avogadro's law, and (*iv*) Ideal gas equation from kinetic gas equation.

Solution :

(*i*) **Boyle's Law:**

The kinetic gas equation is

$$PV = \frac{1}{3}mnC^2 = \frac{2}{3}\times\frac{1}{2}mnC^2 = \frac{2}{3}\times\frac{1}{2}MC^2 \quad (1.16)$$

(Here, $m \times n = M$, the total mass of the gas)

Now, $\frac{1}{2}MC^2$ = kinetic energy of the gas

$$\therefore \quad PV = \frac{2}{3}KE \quad (1.17)$$

According to one of the postulates of kinetic theory of gases kinetic energy of the gas $\propto$ absolute temperature (T), *i.e.*, $KE \propto T$

or $$KE = kT \quad (1.18)$$

where, k is a proportionality constant.

$$\therefore \quad PV = \frac{2}{3}kT \quad (1.19)$$

Thus, at constant temperature,

$$PV = \text{constant} \quad (1.20)$$

This is Boyle's law.

(*ii*) **Charle's Law:**

According to Eq. 1.19,

$$PV = \frac{2}{3}kT$$

$$\Rightarrow \quad \frac{V}{T} = \frac{2}{3}k\times\frac{1}{P} \quad (1.21)$$

Thus, when pressure of the gas is kept constant

$$\frac{V}{T} = \text{constant} \quad (1.22)$$

This is Charle's law.

(*iii*) **Avogadro's Law:**

For any two gases, the kinetic gas equation may be written as

$$p_1 v_1 = \frac{2}{3}\times\frac{1}{2}m_1 n_1 c_1^2 \quad (1.23)$$

$$p_2 v_2 = \frac{2}{3}\times\frac{1}{2}m_2 n_2 c_2^2 \quad (1.24)$$

When the pressures and volumes of two gases are the same, we say

$$p_1 = p_2 \quad \text{and} \quad v_1 = v_2$$

$$\therefore \quad \frac{1}{2} m_1 n_1 c_1^2 = \frac{1}{2} m_2 n_2 c_2^2 \tag{1.25}$$

If these two gases are also at the same temperature, the mean molecular kinetic energy of both the gases would also be the same. Hence,

$$\therefore \quad \frac{1}{2} m_1 c_1^2 = \frac{1}{2} m_2 c_2^2 \tag{1.26}$$

From Eqs. 1.25 and 1.26, we have

$$n_1 = n_2 \tag{1.27}$$

Thus, equal volumes of all gases under the same conditions of pressure and temperature contain equal number of molecules. This is Avogadro's law.

(*iv*) **Ideal Gas Equation:**

From Boyle's law, Charle's law and Avogadro's law, the volume of a gas

$$V \propto \frac{1}{P} \qquad \text{(at constant } T \text{ and } n\text{, Boyle's law)}$$

$$V \propto T \qquad \text{(at constant } P \text{ and } n\text{, Charle's law)}$$

$$V \propto n \qquad \text{(at constant } T \text{ and } P\text{, Avogadro's law)}$$

Combining the three laws, we have

$$V \propto \frac{nT}{P}$$

$$= R\left(\frac{nT}{P}\right)$$

$$\Rightarrow \quad PV = nRT \tag{1.28}$$

Here, R is a proportionality constant, called gas constant, and Eq. 1.28 represents *Ideal gas equation*.

1.2 GAS CONSTANT (*R*)

We know that 1 mol of an ideal gas occupies a volume of 22.414 l at N.T.P. Hence,

$$R = \frac{PV}{nT} = \frac{(1\text{ atm})(22.414\text{ l})}{(1\text{ mol})(273.15\text{ K})}$$

$$\Rightarrow \quad R = 0.08206 \text{ l atm K}^{-1}\text{ mol}^{-1}$$

In S.I. units, 1 atm = 1.01325×10^5 Nm^{-2}

$$\text{Hence,} \quad R = \frac{PV}{nT} = \left(1.01325 \times 10^5 \text{ Nm}^{-2}\right) \frac{\left(22.414 \times 10^{-3} \text{ m}^3\right)}{273.15 \text{ K}}$$

$$\Rightarrow \quad R = 8.314 \text{ N m K}^{-1} \text{ mol}$$
$$= 8.314 \text{ J K}^{-1} \text{ mol (J = Nm)}$$

Problem 2. Derive (*i*) Graham's law of diffusion (*ii*) Dalton's law of partial pressure and (*iii*) Translational kinetic energy of a molecule.

Solution:

(*i*) **Graham's Law of Diffusion:**

As per kinetic gas equation,

$$PV = \frac{1}{3} mnC^2$$

$$\Rightarrow \quad C^2 = \frac{3PV}{mn} \tag{1.29}$$

Since mn is the total mass of the gas molecules present in volume V, the density of the gas (ρ) can be given as:

$$\rho = \frac{mn}{V} \tag{1.30}$$

$$\text{Hence} \quad C^2 = \frac{3P}{\rho} \tag{1.31}$$

$$\Rightarrow \quad C_{\text{rms.}} = \sqrt{C^2} = \sqrt{\frac{3P}{\rho}} \tag{1.32}$$

As C_{rms} is related to average velocity ($<C>$), as

$$C_{\text{rms}} = 1.088 \; < C >, \tag{1.33}$$

$$\text{we have} < C > = \frac{1}{1.088} \sqrt{\frac{3P}{\rho}} \tag{1.34}$$

Thus, at constant pressure,

$$< C > \propto \frac{1}{\sqrt{\rho}} \tag{1.35}$$

As the rate of diffusion of a gas is governed by the average velocity ($< C >$), we say

Rate of diffusion (r) = A constant $\times < C >$

$$= \text{A constant} \times \frac{1}{\sqrt{\rho}} \tag{1.36}$$

Hence, if two gases have their densities as ρ_1 and ρ_2, the ratio of rates of their diffusion can be given as:

$$\frac{r_1}{r_2} = \sqrt{\frac{\rho_2}{\rho_1}} \tag{1.37}$$

Thus, the rate of diffusion of a gas is inversely proportional to the square root of its density. This is **Graham's law of diffusion.**

For 1 mol of the gas, we write the Eq. 1.29 as

$$\Rightarrow \quad C^2 = \frac{3PV_m}{M} \tag{1.38}$$

$$\Rightarrow \quad \sqrt{C^2}\ (= C_{rms}) = \sqrt{\frac{3PV_m}{M}} \tag{1.39}$$

where, M is the molecular weight of the gas and V_m the molar volume.

Thus, at constant pressure, the rate of diffusion of a gas $(r) \propto \frac{1}{\sqrt{M}}$

$$\Rightarrow \quad \frac{r_1}{r_2} = \sqrt{\frac{M_2}{M_1}} \tag{1.40}$$

This is yet another way of expressing Graham's law of diffusion.

(*ii*) **Dalton's Law of Partial Pressure:**

As per kinetic gas equation,

$$PV = \frac{1}{3} mnC^2 = \frac{2}{3} \times \frac{1}{2} mC^2 \times n \tag{1.41}$$

Here, $\frac{1}{2} mC^2 = \varepsilon$ = average molecular kinetic energy (1.42)

$$\therefore \quad P = \frac{2}{3} \varepsilon \frac{n}{V} = \frac{2}{3} \varepsilon N \tag{1.43}$$

where, N is the number of gas molecules per unit volume.

Thus, for two different gases, we have

$$P_1 = \frac{2}{3} \varepsilon_1 N_1 \quad \text{and} \quad P_2 = \frac{2}{3} \varepsilon_2 N_2 \tag{1.44}$$

When these two gases are present at the same temperature, their average molecular kinetic energy will be same, say (ε)

i.e., $\varepsilon_1 = \varepsilon_2 = \varepsilon$ (1.45)

When the molecules are non-interacting and independent of each other, the total pressure (P) exerted by the two gases taken together may be given as

$$P = \left(\frac{2}{3}\varepsilon_1 N_1 + \frac{2}{3}\varepsilon_2 N_2\right) \tag{1.46}$$

Thus, $$P = (P_1 + P_2) \tag{1.47}$$

where, P_1 and P_2 are the partial pressures which each gas would exert, if it is allowed to occupy the same volume. This is *Dalton's law of partial pressure.*

(*iii*) Translational Kinetic Energy of a Gas:

Consider one mole of a gas, which will have N_A number (Avogadro's number) of molecules, and the kinetic gas equation would then be

$$PV = \frac{1}{3} m N_A C^2 = \frac{2}{3} \times \frac{1}{2} mC^2 \times N_A = \frac{2}{3}\varepsilon_t \tag{1.48}$$

Here, ε_t is the kinetic energy of translation of 1 mol of the gas.

Now, for one mole of gas, we have

$$PV = RT$$

Hence $$\varepsilon_t = \frac{3}{2} RT \tag{1.49}$$

It is thus observed that the average kinetic energy of translation of an ideal gas is directly proportional to the absolute temperature. This is called Maxwell generalization. As the average kinetic energy of translation of a mole of gas $\propto C^2$ Eq. 1.48, we have

$$C^2 \propto T$$

$$\Rightarrow \quad C \propto \sqrt{T} \tag{1.50}$$

This implies that molecular velocity of a gas is proportional to the square-root of its absolute temperature. Thus, the molecular motion of a gas is basically its thermal motion and, therefore, when $T = 0$ (*i.e.*, at absolute temperature of zero), the kinetic energy would also be zero and thus its thermal motion would completely cease.

The average kinetic energy of translation of one molecule of an ideal gas may be given as:

$$\varepsilon_t = \frac{\varepsilon_t}{N_A} = \frac{3}{2}\frac{RT}{N_A} = \frac{3}{2} kT \tag{1.51}$$

where, $\frac{R}{N_A} = k$, called Boltzmann constant.

Note that $k = 1.38 \times 10^{-23}$ JK^{-1}.

1.2.1 Deviation from Ideal Behaviour of Gases

An ideal gas obeys the gas equation, $PV = nRT$, at all temperatures and pressures. Hence, a plot of $\left(\frac{PV}{RT}\right)$ vs P is expected to be as shown by dotted line in Fig. 1.2.

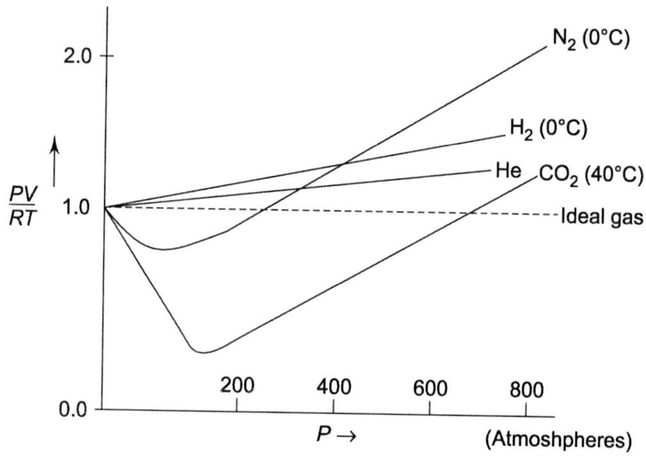

Fig. 1.2 Plot of $\frac{PV}{RT}$ vs. P for some real gases (deviation of real gases from ideal behaviour).

However, for most of the gases, this behaviour is not observed. Thus, for hydrogen and helium, the product (PV) is always greater than expected for ideal gas at all pressures. On the other hand, for gases like CO_2 or N_2, the product (PV) first decreases from ideal gas point, passes through a minimum value and then increase above the value for ideal gas at higher pressures. This deviation of (PV) values of real gases from ideal gas behaviour can be expressed by a **compressibility factor (Z).**

$$Z = \frac{(PV) \text{ observed for Real gas}}{(PV) \text{ for Ideal gas}} = \frac{PV}{nRT} \tag{1.52}$$

For an ideal gas, $Z = 1$ under all conditions of pressures and temperatures. Deviation from a value of unity of the compressibility factor is a measure of the non-ideality of the gas. At low pressures and 40°C temperature, carbon dioxide is more compressible than an ideal gas, so that product (PV) is less than (PV ideal) and thus $Z < 1$. In fact over a range of low pressures, the attractive forces among molecules are dominant and favour compression of gas. However, at high pressure range, when $Z > 1$, the product PV gets a higher value than PV ideal.

If plots of $\left(\frac{PV}{RT}\right)$ vs P for real gases (*e.g.*, nitrogen) are carefully studied at varying temperatures (between – 70 and 50°C) the factor Z almost attains unity value under a reasonable range of pressures (approximately near 100 atm). Thus, the product PV becomes constant or Boyle's law is obeyed under such a pressure range and at 50°C for nitrogen. This temperature is called Boyle point or Boyle temperature. The Boyle temperature for hydrogen is –165°C.

1.2.2 Causes of Deviation of Real Gases from Ideal Behaviour

It appears that the following two assumptions of kinetic theory of gases need a fresh approach, so as to explain the deviations from ideality of real gases.

1. The actual volume occupied by the gas molecules is negligible in comparison to the total volume of the gas.
2. The gas molecules do not exert any appreciable force of attraction or repulsion between themselves.

The gas molecules themselves occupy a certain volume and this cannot be neglected in comparison to the total volume of gas, especially under conditions of higher pressures or low temperatures.

Similarly, under conditions of higher pressure or low temperatures, when molecules lie quite close to each other, their intermolecular forces of attraction are so appreciable that these can no longer be neglected.

Thus, certain corrections are to be made in these faulty assumptions or postulates of kinetic theory of gases so that the behaviour of real gases is correctly represented by mathematical equations and explained properly.

1.3 THE VAN DER WAALS EQUATION OF STATE

J.D. van der Waals obtained an equation for the real gases by introducing following two correction terms :

(*i*) **Correction Term "*b*" for Finite Volume of Gas Molecules:** Van der Waals suggested that gas molecules have a definite finite size. If b is the effective volume of the gas molecules per mole of the gas, and total volume of the gas occupied by n moles of gas is V, then volume available (free space) for molecules to move freely is $(V - nb)$ and not V. This term 'b' is called excluded volume or co-volume and its value per molecule is $= 4 \times$ geometric volume of 1 mol of molecules $(N_A\ V_m)$.

Here, N_A is the Avogadro's number and $V_m = \frac{4}{3}\pi r^3$, where r is the radius of the molecule Fig. 1.3.

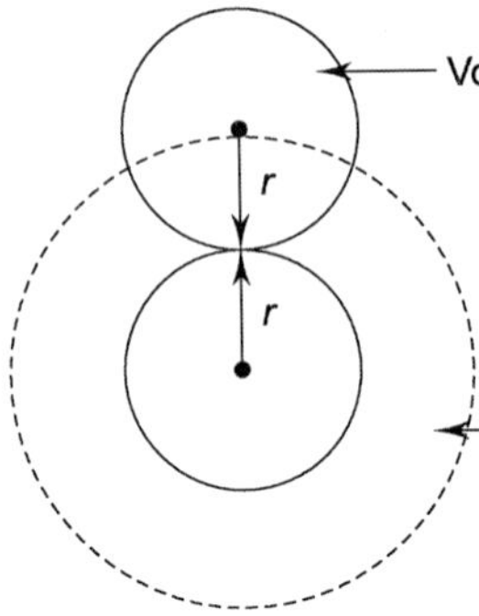

Fig. 1.3 Excluded volume per pair of molecules.

(*ii*) **Pressure Correction ("*a*") – Intermolecular Forces of Attraction:** A gas molecule approaching to strike the wall (*W*) of a container Fig. 1.4 experiences an inward pull due to intermolecular forces of attraction exerted by the molecules surrounding it. On the other hand, a gas molecule present in the bulk of the gas (*M*), being uniformly attracted by other molecules, does not experience any resultant pull from any direction. Hence, the gas molecule approaching the wall of the container would be striking it with a bit slow velocity and would exert a lower pressure. Thus, the ideal pressure term should be $(P + p)$ and not just P. This term p is the "correction factor" due to inward pull on the gas molecules approaching the wall, and depend upon:

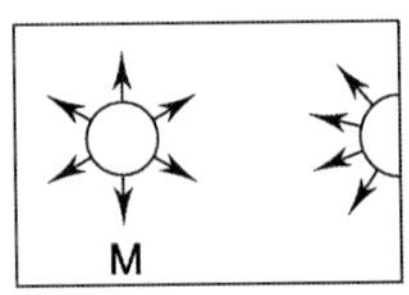

Fig. 1.4 Intermolecular attraction.

1. Number of molecules surrounding a molecule of gas, *i.e.*, upon the density of the gas.
2. Number of molecules striking the wall at any given time which again depends on the density of gas.

Thus, the "correction factor" for a given mass of a gas is inversely proportional to its volume (V)

i.e.,
$$p \propto \frac{1}{V^2} \quad (1.53)$$

or
$$p = \frac{a}{V^2} \quad (1.54)$$

where, a is a constant that depends on nature of the gas.

Thus, incorporating the correction factor, the corrected pressure of a real gas becomes

$$= \left(P + \frac{a}{V^2}\right) \quad (1.55)$$

In view of correction terms (*i*) and (*ii*), the kinetic gas equation $PV = RT$ (for 1 mol of a new gas) is modified as

$$\left(P + \frac{a}{V^2}\right)(V - b) = RT \quad (1.56)$$

This equation is named as the van der Waals equation.

Considering n moles of gas, we have

$$p \propto \frac{n^2}{V^2}$$

or
$$p = a\frac{n^2}{V^2} \quad (1.57)$$

The van der Waals equation for n moles of gas is, therefore, represented as:

$$\left(P + a\frac{n^2}{V^2}\right)(V - nb) = nRT \quad (1.58)$$

The terms "*a*" and "*b*" in the van der Waals equation are the measure of **intermolecular forces of attraction** and **effective volume** of the gas, respectively, and are constant for a given gas.

The van der Waals constants, *i.e.*, a is a measure of '**intermolecular forces of attraction** and b is the **effective volume** of the gas.

For carbon dioxide gas, the value of a is 3.59 dm^6 atm mol^{-2} and that of b is 0.0427 dm^3 mol^{-1}.

Problem 3. Calculate the molecular diameter of CO_2, if it is assumed to be a spherical molecule. Use the value of van der Waals constants given above and take Avogadro's number as 6.02×10^{23}.

Solution:

For 1 mol of gas

$$b = 4\,N_A\,V_m$$

Here
$$N_A = 6.02 \times 10^{23}$$

and
$$V_m = \frac{4}{3}\pi r^3 = \frac{b}{4N_A}$$

If σ is molecular diameter, then

$$r^3 = \left(\frac{\sigma}{2}\right)^3$$

$$\therefore \quad \frac{4}{3}\pi\left(\frac{\sigma}{2}\right)^3 = \frac{b}{4N_4}$$

$$\Rightarrow \quad \sigma^3 = \frac{b}{4N_A} \times \frac{3}{4} \times \frac{8}{\pi} = \frac{0.0427 \times 3}{2 \times 3.143}$$

$$= 33.84 \times 10^{-24}$$

$$= 3.23 \times 10^{-8}\ \text{cm} = 3.23\ \text{Å}.$$

1.3.1 Behaviour of Real Gases and van der Waals Equation

For 1 mol of a real gas, the van der Waals equation is written as

$$\left(P + \frac{a}{V_m^2}\right)(V_m - b) = RT \tag{1.59}$$

$$\Rightarrow \quad P\,V_m = RT + Pb - \frac{a}{V_m} + \frac{ab}{V_m^2} \tag{1.60}$$

Behaviour of the real gas, under different conditions, is discussed below:

(i) At Low Pressure: When P is small, V_m will be sufficiently large and the terms Pb and $\frac{ab}{V_m^2}$ can be ignored, hence

$$P\,V_m = RT - \frac{a}{V_m} \tag{1.61}$$

Thus, the product PV_m is less than RT and the difference goes on increasing with increase in pressure. As such, PV_m values falls in this region, which explains the dip of the isotherm of most of gases Fig. 1.2.

(ii) At High Temperature: At any given pressure, if the temperature is considerably high, V is very large to make the value of $\frac{a}{V_m^2}$ negligibly small. Even 'b' can be neglected in comparison to large V_m. Thus, van der Waals equation approaches the ideal gas equation $PV = RT$. As such, the deviations are less marked at high temperatures.

(iii) At Very High Pressure: The terms $\frac{a}{V_m}$ and $\frac{ab}{V_m^2}$ may be ignored in comparsion to the terms Pb. Hence,

$$PV_{in} = RT + Pb \tag{1.62}$$

Obviously, the term PV_m is now greater than RT by a factor Pb and it goes on increasing with increase of pressure. As such, PV_m value increases after reaching a minimum on increasing the pressure Fig. 1.2.

(*iv*) **Exceptional Behaviour of Hydrogen and Helium:** Hydrogen and helium gas molecules have very small masses and so, the attractive forces acting between these molecules are much smaller. Thus, the correction term $\frac{a}{V_m^2}$ introduced for attraction factor is almost negligible at ordinary temperatures, and the Eq. 1.59 takes the shape

$$P(V_m - b) = RT \tag{1.63}$$

$$\Rightarrow \quad PV_m = RT + Pb \tag{1.64}$$

Thus, PV_m value for these gases increases from RT value with increasing pressure at ordinary temperature. The Boyle's temperature (T_b), when a real gas obeys Boyle's law, is given as

$$T_b = \frac{a}{bR} \tag{1.65}$$

where, a and b are van der Waals constants.

Problem 4. Calculate Boyle's temperature, when the van der Waals constants for CO_2 are:

a = 3.59 dm^6 atm mol^{-2} and

b = 0.0427 dm^3 mol^{-1}

R = 0.08206 dm^3 atm K^{-1} mol^{-1}

Solution:

$$T_b = \frac{a}{bR}$$

$$= \frac{3.59 \text{ dm}^6 \text{ atm mol}^{-2}}{\left(0.0427 \text{ dm}^3 \text{ mol}^{-1}\right)\left(0.08206 \text{ dm}^3 \text{ atm K}^{-1} \text{ mol}^{-1}\right)}$$

T_b = 1025.7 K.

1.4 CRITICAL PHENOMENON: *PV* ISOTHERMS OF REAL GASES

We know that by lowering the temperature of the gas, the kinetic energy of its molecules decreases, and they come close to each other so that the intermolecular forces of attraction acting between them are immensely increased leading to liquefaction of gases into the liquid at a particular temperature. Similar phenomenon also takes place when a high pressure is applied on the gas causing its liquefaction. The temperature at which a gas liquefies by the applied high pressure is called **critical temperature** (T_c), and the pressure corresponding to the critical temperature is called **critical pressure** (P_c). The volume occupied by 1 mol of gas at its critical temperature and critical pressure is called **critical volume** (V_c).

Critical temperature of carbon dioxide is 31.1°C, when its critical pressure is 72.9 atm and the critical volume is 940.0 ml mol^{-1}. The characteristic constants T_c, P_c and V_c of a gas are known as critical constants.

T. Andrews (1869) carried out measurements of pressure and volume data of carbon dioxide gas at a series of temperatures. Plot of such *P–V* data (isotherms) of CO_2 over a range of 48.1° C and 13.1°C are presented in Fig. 1.5.

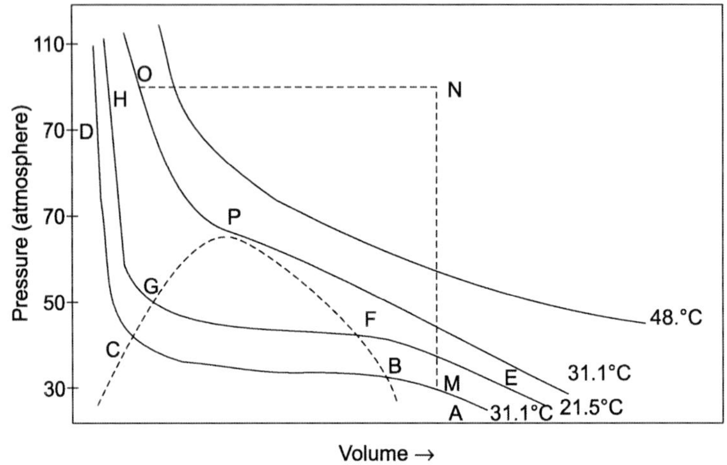

Fig. 1.5 *P-V* isotherm of CO_2.

It can be seen that at a temperature as high as 48.1°C, the isotherm obtained is quite a smooth curve obeying Boyle's law. On lowering the temperature, a distinct deformity appears in the isotherm at 31.1°C, when increasing the pressure along *FP* causes a decrease in volume. At point *P* (72.8 atm), the CO_2 gas appears in liquid state. Below this temperature, *e.g.,* at 21.5°C, the isotherm (*EFGH*) shows gaseous behaviour at 100 atm pressure (from *E* to *F*). However, a sudden discontinuity at *F* appears when pressure is increased. The gas shows a sudden change in state and is liquefied, and an appreciable change in volume is caused.

It has been concluded by Andrews that carbon dioxide cannot be liquefied above 31.1°C even by applying higher pressure. The isotherm at this temperature is called the *critical isotherm* and the corresponding temperature for carbon dioxide is referred to as **critical temperature** (T_c) and the pressure as **critical pressure** (P_c). The volume occupied by 1 mole of carbondioxide at T_c (31.1°C) and P_c (72.8 atm) is 940.0 ml and is known as its critical volume (V_c).

1.5 CONTINUITY OF STATES

Consider the isotherm of carbon dioxide at 31.1°C of Fig. 1.5. At point *P*, the critical point, gas is transformed into liquid without any sudden change in properties, *e.g.,* density, refractive index, *etc.*, of the system. Under this condition, the gas–liquid transition is continuous, *i.e.*, at any given time, only a single phase exists. The liquid phase gradually develops in the gas and the gas phase disappears. Below this temperature, we have a horizontal portion in all the isotherms wherein both the phases, *i.e.*, liquid and vapour co-exist. Joining the two ends of the horizontal portion of the isotherm, a dotted parabolic curve is obtained. Within the area of this curve both phases, *i.e.*, liquid and gas co-exist.

Let us take a point *M* Fig. 1.5, correspondingly the substance (CO_2) is in gaseous state at this point. If this gas is heated so that system of gas moves along MN by application of higher pressures. On reaching the point *N*, the gas is now kept at constant pressure and allowed to decrease its temperature, thereby, a decrease of its volume along NO take place till it reaches to point *O* on the curve on the critical isotherm and it then liquefies. This liquefaction of gas takes place along the isotherm, without any discontinuity, *i.e.*, system at every time has only single phase–either gas or liquid. Thus, we say that gas–liquid transition is gradual and continuous or it can be referred to as continuity of state.

1.6 ISOTHERMS BASED ON VAN DER WAALS EQUATION

For 1 mol of a gas, van der Waals equation is given as

$$\left(P+\frac{a}{V_{\mathrm{m}}^2}\right)\left(V_{\mathrm{m}}-b\right)=RT \tag{1.66}$$

where, V_{m} is the molar volume.

The above equation can be expanded to take the form

$$PV_{\mathrm{m}}-P\times b+\frac{a}{V_{\mathrm{m}}}-\frac{ab}{V_{\mathrm{m}}^2}-RT=0 \tag{1.67}$$

Multiplying Eq. 1.67 by V_{m}^2, dividing by P and then rearranging the resulting expression in decreasing powers of V_{m}, we get

$$V_{\mathrm{m}}^3-\left(b+\frac{RT}{P}\right)V_{\mathrm{m}}^2+\frac{aV_{\mathrm{m}}}{P}-\frac{ab}{P}=0 \tag{1.68}$$

This is a cubic equation in V_{m} and for a single set of values of pressure and temperature for 1 mole of gas, the Eq. 1.68 will provide three values of volume corresponding to the three roots of this cubic equation. Thus, this equation, which is another form of van der Waals equation can be used to draw *P–V* curves (isotherms) at the desired temperature. Such isotherms plotted for CO_2 Fig. 1.6 , using values of pressures and van der Waals constants, gave three different values of *V*. Isotherms plotted upto critical temperature (31.1°C) are similar to those Fig. 1.6 obtained by using experimental values of Andrews, except for the wave-like portion *RQ* and *NO*. These are representing super-saturated vapour and supercooled liquid, respectively. In the isotherms obtained for higher temperatures, the loops are getting closer till at the critical temperature the three values of *V*, corresponding to points *R*, *P* and *N* of the isotherm, merge into a single value (critical volume) at point *X*. Thus, van der Waals equation postulates that

$$V_{\mathrm{m}}=V_{\mathrm{c}} \tag{1.69}$$

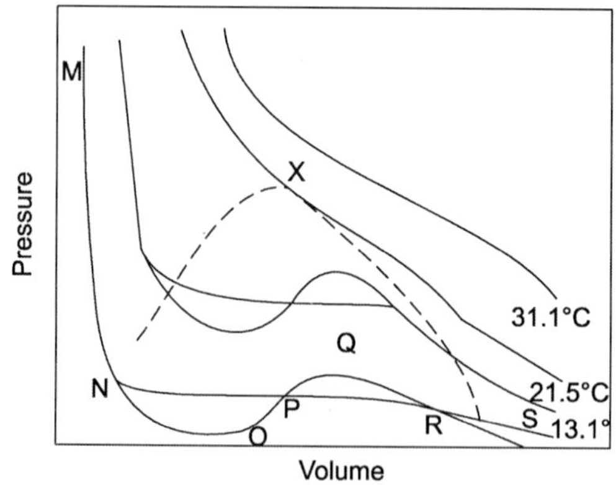

Fig. 1.6 *P-V* isotherms based on van der Waals equation.

1.7 RELATIONSHIP BETWEEN CRITICAL CONSTANTS AND VAN DER WAALS CONSTANTS

From Eq. 1.69

$$(V_m - V_c)^3 = 0 \tag{1.70}$$

Expanding and rewritting Eq. 1.71 in decreasing powers of V_m, we get

$$(V_m)^3 - 3V_c\,(V_m)^3 + 3\,(V_c)^2\,V_m - (V_c)^3 = 0 \tag{1.71}$$

Equation (1.71) must be identical with the van der Waals Eq. 1.68) which, for 1 mole of gas at critical temperature and pressure, can be written as:

$$V_m^3 - \left(b + R\frac{T_c}{P_c}\right)(V_m)^2 + \frac{a}{P_c}V_m - \frac{ab}{P_c} = 0 \tag{1.72}$$

Comparing the coefficients of equal powers of V_m in the Eqs.1.71 and 1.72, we get

$$3V_c = \frac{RT_c}{P_c} + b \tag{1.73}$$

$$3V_c^2 = \frac{a}{P_c} \tag{1.74}$$

and

$$V_c^3 = \frac{a}{P_c}b \tag{1.75}$$

Dividing Eq. 1.75 by Eq. 1.74, one gets

$$\frac{1}{3}V_c = b \text{ or } V_c = 3b \tag{1.76}$$

Substituting this value of V_c in Eq. 1.74,

$$3\,(3b)^2 = \frac{a}{P_c}$$

$$\Rightarrow \qquad P_c = \frac{a}{27b^2} \qquad (1.77)$$

Inserting these values of P_c and V_c in Eq. 1.73,

$$T_c = \frac{8a}{27\,Rb} \qquad (1.78)$$

Thus, knowing the critical constants of a gas, we can obtain its van der Waals constants and *vice versa.*

Normal boiling point of a liquid is approximately two-third of its critical temperature, *i.e.*, $\left(\frac{T_b}{T_c}\right) \approx \frac{2}{3}$.

Problem 5. The van der Waals constants of a gas are $a = 3.61$ atm dm^6 mol^{-2} and $b = 0.0428$ dm^3 mol^{-1}. Calculate its critical constants.

Solution:

$$V_c = 3b = 3 \times 0.0428 \text{ dm}^3 \text{ mol}^{-1}$$
$$= 0.1284 \text{ dm}^3 \text{ mol}^{-1}$$

$$P_c = \frac{a}{27b^2} = \frac{3.61 \text{ dm}^6 \text{ atm mol}^{-2}}{27 \times \left(0.0428 \text{ dm}^3 \text{ mol}^{-1}\right)^2}$$

$$\therefore \qquad = 72.9 \text{ atm}$$

and
$$T_c = \frac{8a}{27\,Rb}$$

$$= \frac{8 \times 3.61 \text{ dm}^3 \text{ mol}^{-1}}{27 \times \left(0.8206 \text{ dm}^3 \text{ atm K}^{-1} \text{ mol}^{-1}\right) \times \left(0.0428 \text{ dm}^3 \text{ mol}^{-1}\right)}$$
$$= 304 \text{ K.}$$

1.7.1 Determination of Critical Constants, Measurement of T_c and P_c

The important observation that has been applied in determining P_c and T_c is, that the "surface of separation (*i.e.*, the meniscus) between a liquid and its vapour phase disappears at critical temperature".

The experimental liquid is taken in the bulb *A* of the apparatus Fig. 1.7. This bulb is being surrounded by a jacket (J) through which hot/cold water of desired temperature is circulated. The meniscus of the liquid is observed and the pressure corresponding to the temperature of disappearance of liquid meniscus is read through the manometer (M). The bulb is now cooled by circulating the cold water through the jacket and temperature of reappearance of liquid meniscus is noted along with measurement of pressure corresponding to the temperature of reappearance. The mean of these two temperatures gives the critical temperature (T_c) of the liquid/vapour enclosed in the bulb. Value of the corresponding critical pressure (P_c) is noted through the manometer.

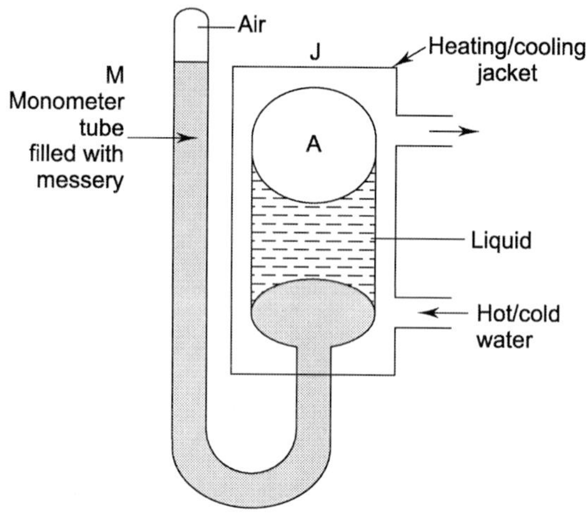

Fig. 1.7 Determination of critical temperature (T_c) and pressure (P_c).

1.7.2 Measurement of Critical Volume (V_c)

The law proposed by Cailletet and Mathias (1886) states that the mean of the densities of any substance in the liquid state and its saturated vapour (at the same temperature) varies linearly with temperature. This basic concept is employed in determining the critical volume of the substance (V_c). Hence, the densities of the substance in the liquid form and that of its saturated vapours are measured as a function of temperature. The data are then plotted as shown in Fig. (1.8).

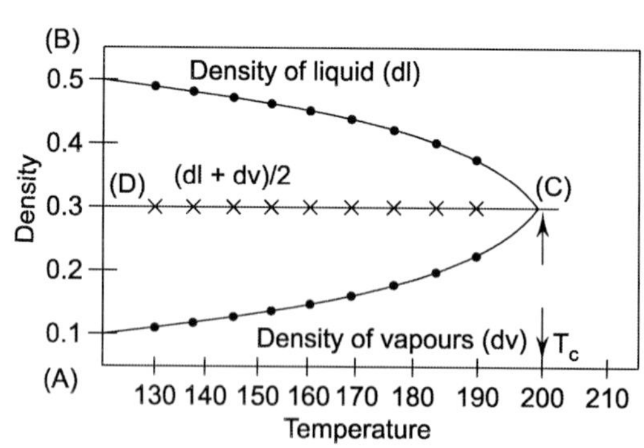

Fig. 1.8 Plot of densities of liquid and the corresponding saturated vapours against temperature (in Kelvin).

As the temperature is raised, the density of liquid decreases along BC and that of its saturated vapour increases along AC. The two densities approach each other giving a parabolic curve BCA. The mean of the two densities is then plotted against varying temperature. A straight line (DC) thus obtained is allowed to intersect the parabolic curve at point C, which gives the value of critical density (ρ_c) at T_c. The critical volume (V_c) is then calculated as

$$V_c = \frac{M}{\rho_c} \tag{1.79}$$

where, M is the molar mass of the substance.

1.8 THE LAW OF CORRESPONDING STATES

It is convenient to express pressure, temperature and volume of any gas as a fraction related to the corresponding critical constant. Thus, we get a set of new variables called **reduced variables** *e.g.,* the reduced pressure (π), the reduced temperature (θ) and the reduced volume (ϕ) and express them as

$$\pi = \frac{P}{P_c}\ ;\ \phi = \frac{V}{V_c} \quad \text{and} \quad \theta = \frac{T}{T_c} \tag{1.80}$$

Substituting these values of P, V and T in van der Waal's Eq. 1.56), we get

$$\left[\pi P_c + \frac{a}{(\phi V_c)^2}\right]\left[\phi V_c - b\right] = R \cdot \theta \cdot T_c \tag{1.81}$$

Substituting the values of P_c, V_c and T_c in terms of van der Waals constants [Refer Eqs.1.76, 1.77–1.78], we have

$$\left[\pi \frac{a}{27b^2} + \frac{a}{(\phi\, 3b)^2}\right]\left[\phi\, 3b - b\right] = R\,\theta \frac{8a}{27\,Rb}$$

$$\Rightarrow \qquad \left(\pi + \frac{3}{\phi^2}\right)(3\phi - 1) = 8\theta. \tag{1.82}$$

This equation is termed the **reduced equation of state**. As this equation does not involve van der Waals constants a and b, it can be applied to all gases.

It can be concluded from Eq. 1.82 that *If any two or more substances have the same reduced pressure, and are at equal reduced temperatures, their reduced volumes must also be equal.* This conclusion is known as the **law of corresponding states**. Substances having the same reduced pressure, the same reduced temperature and also the same reduced volume are said to be in the **corresponding state**.

1.9 MOLECULAR VELOCITIES : ROOT MEAN SQUARE, AVERAGE AND MOST PROBABLE VELOCITIES

1.9.1 Most Probable Velocity (C_{mp})

It is the velocity with which the maximum number of gas molecules are moving at a particular temperature. It is given as

$$C_{mp} = \left[\frac{2RT}{M}\right]^{\frac{1}{2}} \tag{1.83}$$

where, M is the molecular weight of the gas molecule.

1.9.2 Average Velocity ($<C>$)

It is the arithmetic mean of the velocities with which different molecules of a gas are moving at a given temperature.

$$<C> = \frac{v_1 + v_2 + v_3 + \cdots + v_n}{n} \tag{1.84}$$

The average velocity is also expressed as

$$<C> = \left[\frac{8RT}{\pi M}\right]^{\frac{1}{2}} \tag{1.85}$$

1.9.3 Root Mean Square Velocity (C_{rms})

It is the square root of the mean of the squares of different velocities with which the gas molecules are moving at a given temperature. It is assumed that amongst a population of n molecules, if n_1 molecules possess molecular velocity v_1, n_2 molecules possess molecular velocity v_2 and so on, then,

$$C_{rms} = \left(\frac{v_1^2 + v_2^2 + \cdots + v_n^2}{n}\right)^{\frac{1}{2}} \tag{1.86}$$

The root mean square velocity is also expressed as

$$C_{rms} = \left(\frac{3RT}{M}\right)^{\frac{1}{2}} \tag{1.87}$$

where, M is the gram molecular weight of gas molecules.

1.10 MAXWELL'S DISTRIBUTION OF MOLECULAR VELOCITIES

The molecules of gas at any temperature are in continuous motion. They collide with one another and during these collisions their molecular velocities get affected. Thus, it is not possible to assign a definite value of molecular velocity to any particular molecule as they continuously exchange energy during the collisions and their velocities keep on changing every 10^{-9} s. It is, therefore, appropriate to talk about distribution of molecules over possible velocity ranges from near zero to quite high values. Boltzmann predicted that the probability for a molecule to have an energy ε was proportional to $e^{-\varepsilon/kT}$. J.C. Maxwell, using theory of probability, could arrive at the result that the fraction of molecules having velocity between C and $C + dC$ is expresed as

$$\frac{dN}{N} = 4\pi\left(\frac{M}{2\pi kT}\right)^{\frac{3}{2}} C^2 \exp\left[-\frac{MC^2}{2RT}\right] dC \tag{1.88}$$

Here, M is the molecular weight and N the total number of gas molecules. This expression is known as *Maxwell's distribution of molecular velocities.* This is better represented as a plot of the fraction $\left[\frac{1}{N}\left(\frac{dN}{dC}\right)\right]$ taken on y-axis vs C taken on x-axis. A typical Maxwell's plot at two different temperatures is shown in Fig. 1.9.

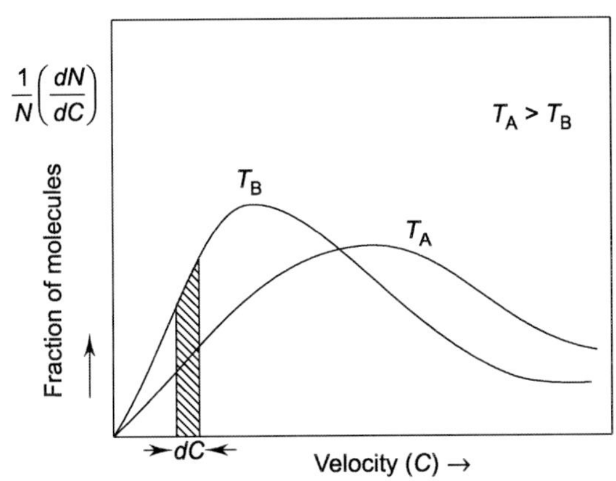

Fig. 1.9 Maxwell's distribution of molecular velocity.

The important features of the Maxwell's plot can be summarised as follows:

1. The dotted area enclosed by dC (x-axis) and the ordinate would be given by $\frac{dN}{N}$, *i.e.*, the fraction of the total population of molecules having velocities C and $[C + dC]$. Hence, it is found that the total number of molecules would be given by the area of the curve enclosed (at a particular temperature) between length of x-axis and height of ordinate at the velocity value C.
2. It is further observed that a very small fraction of the total molecules move either with higher or lower values of velocity. The maxima of any curve provides the fraction of the total population of molecules moving with the most probable velocity (C_{mp}).

3. On increasing the temperature, the nature of the curve remains unaltered. Only the maxima of the curve gets flat and shifts to the right, thereby indicating that at higher temperatures, a wider population possesses higher velocity, *i.e.*, increase of temperature favours an increase in the population of molecules possessing most probable velocity.

Problem 6. Calculate the most probable velocity, the average velocity and the root mean square velocity for carbon dioxide of 298 K.

Solution:

We know that molar mass of CO_2= 44 g mol^{-1} = 44 × 10^{-3} kg mol^{-1}.

Now, the most probable velocity (C_{mp}) = $\left(\frac{2RT}{M}\right)^{\frac{1}{2}}$

$$= \left[\frac{\left(2\times 8.314\ JK^{-1}\ mol^{-1}\right)\left(298\ K\right)}{44\times 10^{-3}\ kg\ mol^{-1}}\right]^{\frac{1}{2}}$$

$$= 335.6\ ms^{-1}.$$

The average velocity (< C >) = $\left(\frac{8RT}{\pi M}\right)^{\frac{1}{2}}$

$$= \left[\frac{8\times 8.314\ JK^{-1}\ mol^{-1}\times 298\ K}{3.14\times 44\times 10^{-3}\ kg\ mol^{-1}}\right]^{\frac{1}{2}}$$

$$= 378.8\ ms^{-1}$$

And, the root mean square velocity (C_{rms})

$$= \left(\frac{3RT}{M}\right)^{\frac{1}{2}}$$

$$= \left[\frac{3\times 8.314\ JK^{-1}\ mol^{-1}\times 298\ K}{44\times 10^{-3}\ Kg\ mol^{-1}}\right]^{\frac{1}{2}}$$

$$= 411\ ms^{-1}.$$

1.11 COLLISION NUMBER, MEAN FREE PATH AND COLLISION DIAMETER

Kinetic theory of gas predicts that the gas molecules do collide with themselves. When two gas molecules, assumed to be rigid spheres of finite size, approach quite close to each other, their mutual repulsion, by virtue of electronic and nuclear repulsive forces, causes them to drift away from each other and a molecular collision between them is said to have taken place. Under such circumstances, the closest distance between the centre of two colliding molecules is called the **collision diameter** (σ) Fig. 1.10.

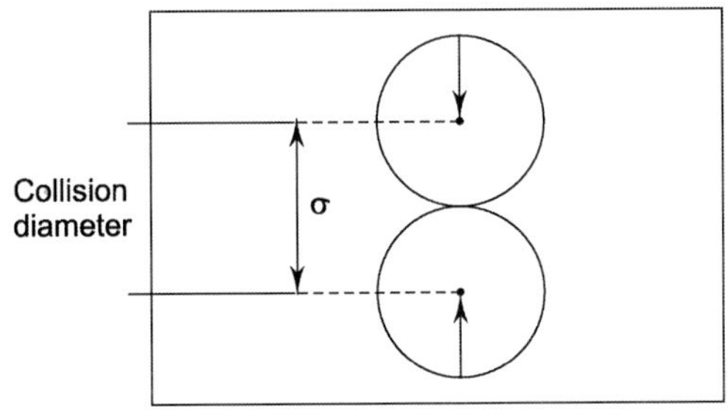

Fig. 1.10 Head-on collision of two gas molecules.

1.11.1 Collision Number

Collision number (Z_1) is the number of collisions which a single molecule will have in unit time through other molecules present in unit volume of the gas. It is expresed as

$$Z_1 = \sqrt{2}\,\pi\sigma^2 < C > \rho \quad (1.89)$$

where, $< C >$ is the average velocity of gas molecules and ρ the gas density, *i.e.*, the number of gas molecules present in unit volume of the gas. Now substituting the value of average velocity from Eq. 1.85, and taking the gas density $(\rho) = \frac{P}{kT}$, we have

(1.90)

$$Z_1 = \sqrt{2}\,\pi\,\sigma^2 \left(\frac{8RT}{\pi M}\right)^{\frac{1}{2}} \frac{P}{kT} \quad (1.91)$$

It is evident that each collision involves two molecules. Hence, the total number of collisions occurring among the same molecules in unit time per unit volume of gas would be

$$Z_{11} = \frac{1}{2}\left(\sqrt{2}\,\pi\sigma^2 < C > \rho^2\right)$$

Since $\qquad \rho = \frac{P}{kT} \quad <C> \left[\frac{8RT}{\pi M}\right]^{\frac{1}{2}}$

$$= \frac{1}{\sqrt{2}}\left\{\pi\sigma^2\left[\frac{8RT}{\pi M}\right]^{\frac{1}{2}} \frac{P^2}{k^2T^2}\right\}$$

$$= 2\sigma^2\left(\frac{\pi RT}{M}\right)^{\frac{1}{2}}\left(\frac{P}{kT}\right)^2 \quad (1.92)$$

1.11.2 Mean Free Path (λ)

It is the average distance travelled by a gas molecule between the two successive collisions. Hence,

$$\lambda = \frac{\text{The distance travelled by a gas molecule in a definite time interval}}{\text{Number of collisions taking place by this molecule in the same time interval}} \tag{1.93}$$

i.e.,
$$\lambda = \frac{<C>}{Z_1}$$

$$= \left(\frac{8RT}{\pi M}\right)^{\frac{1}{2}} \times \frac{1}{\sqrt{2}\pi\sigma^2\left(\frac{8RT}{\pi M}\right)^{\frac{1}{2}}} \times \frac{kT}{P}$$

$$= \frac{kT}{\sqrt{2}\pi\sigma^2 P} \tag{1.94}$$

Thus, it can be concluded that the mean free path of a gas molecule is inversely proportional to its pressure at constant temperature.

Problem 7. Calculate the mean free path of nitrogen molecule, whose diameter is 3.74×10^{-10} m, and it is kept at 0.658 atm pressure and 27°C temperature.

Solution: The mean free path (λ) is given by the expression

$$\lambda = \frac{kT}{\sqrt{2}\pi\sigma^2 P}$$

$$= \frac{RT}{\sqrt{2}\pi\sigma^2 P N} \quad \left(\text{since } k = \frac{R}{N}\right)$$

$$= \frac{0.0821\,\text{atm}^{-1}\,K^{-1}\,\text{mol}^{-1} \times 300\,K}{1.4142 \times 3.14 \times \left(3.74 \times 10^{-10}\,\text{m}\right)^2 \times 0.658\,\text{atm} \times 6.02 \times 10^{23}}$$

$$= 0.032 \times 10^{-10}\,\text{m}.$$

1.12 LIQUEFACTION OF GASES (BASED ON JOULE–THOMSON EFFECT)

It has been observed under the discussions of critical phenomena that a gas can be liquefied while cooling it below its critical temperature and compressing it at critical pressure.

1.12.1 Linde's Method (Application of Joule–Thomson Effect)

Famous English physicists James Joule and Lord Kelvin had stated that cooling effect is observed when a compressed gas is allowed to expand through a small porus plug or orifice, into a region of low pressure. Thus, ordinary gases can be cooled in Joule–Thomson (J.T.) expansion and can be easily liquefied. Linde's apparatus for liquefying air is shown in Fig. 1.11.

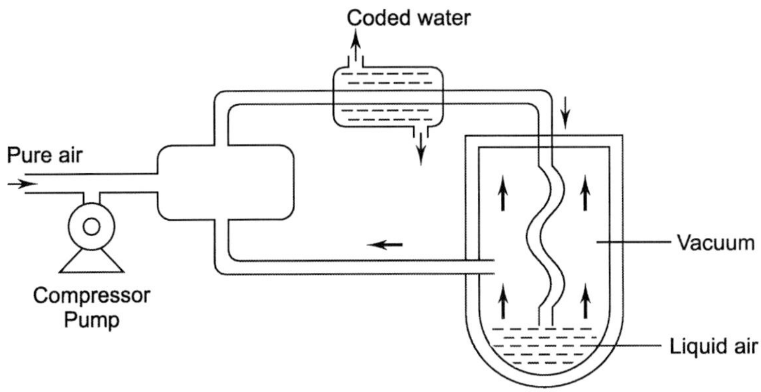

Fig. 1.11 Linde's apparatus for liquefication of air.

Pure air is allowed to be compressed to very high pressure of approximately 200 atm, using powerful compression pump. This is then further cooled while passing through the heat exchanger. When such a highly compressed cooled gas is expanded to a very low pressure region by passing through a fine nozzle its temperature is remarkably lowered due to sudden expansion of air as per J.T. Effect. This cyclic process, when repeated several times by re-circulating the cooled air through compression and J.T. expansion, causes its liquefaction and the liquefied air gets collected at the bottom of the jacket.

Objective Type Questions

1. The change of momentum of a perfectly ideal gas molecule moving with a velocity v_x, after hitting the wall, would be
   (*a*) mv_x (*b*) $2mv_x$
   (*c*) zero (*d*) $-mv_x$
2. What would be the mean free path of a gas at 6 atm, if its value at 1 atm is λ?
   (*a*) $6l$ (*b*) $\frac{\lambda^2}{6}$
   (*c*) $\frac{\lambda}{6}$ (*d*) $12l$
3. Kinetic theory of gases predicts that the average kinetic energy
   (*a*) is zero at 0°C (*b*) decreases with rise in temperature
   (*c*) is always constant (*d*) is proportional to absolute temperature

4. The correct van der Waals equation for n moles of gas is

(*a*) $\left(P+\frac{a}{V^2}\right)(V-b) = nRT$

(*b*) $\left(P+a\frac{n^2}{V^2}\right)(V-nb) = nRT$

(*c*) $\left(P+\frac{a}{V^2}\right)(V-nb) = nRT$

(*d*) $\left(P+\frac{a^2n}{V^2}\right)(V-nb) = nRT$

5. The correct statement out of the following is:

(*a*) A gas can be liquified at its reduced temperature

(*b*) At critical temperature, the surface of separation between liquid and vapour phases disappears

(*c*) A gas can be liquefied by applying a higher pressure without cooling it to critical temperature

(*d*) Critical temperature of oxygen gas is 31.1°C

6. Critical volume V_c and reduced volume (f) are related as:

(*a*) $V \times V_c = \phi$ (*b*) $V \times \phi = V_c$

(*c*) $V_c \times \phi = V$ (*d*) $V \times V_c \times f = 1$

7. van der Waals constant b for CO_2 is 0.042 and for CCl_4, it is 0.1383. The excluded volume of CCl_4 is:

(*a*) greater than that of CO_2 (*b*) less than CO_2

(*c*) equal to that of CO_2 (*d*) not related to the value of b

8. van der Waals constant b for NH_3 and Cl_2 are 0.0371 and 0.0562 respectively. The critical volume of ammonia is:

(*a*) greater than that of Cl_2 (*b*) less than that of Cl_2

(*c*) equal to that of Cl_2 (*d*) not related to value of b

9. The critical temperature and pressure of a gas are 5.2 K and 2.25 atm, respectively. Its critical volume is:

(*a*) 3 l mol^{-1} (*b*) 0.03 l mol^{-1}

(*c*) $\frac{5.2}{2.25}$ l mol^{-1} (*d*) 2.03 l mol^{-1}

10. An equation not involving constants a and b is:

(*a*) van der Waals equation (*b*) Reduced equation of state

(*c*) Gas equation for real gases (*d*) Boyle's temperature equation

ANSWERS

1. (*d*)	**2.** (*b*)	**3.** (*d*)	**4.** (*b*)	**5.** (*b*)
6. (*c*)	**7.** (*a*)	**8.** (*b*)	**9.** (*b*)	**10.** (*b*)

Short Answer Type Questions

1. Give postulates of the kinetic theory of gases.

2. Explain r.m.s. velocity of gas molecules.

3. Describe collision frequency.

4. What is meant by mean free path ?

5. Explain what is meant by the most probable velocity?

6. Describe the law of corresponding states.

7. Find out the pressure exerted by 66 g of CO_2 at 31°C, occupying a volume of 1.98 l. Given that; $a = 3.61$ atm l^2 mol^{-2}, $b = 0.0428$ l.

[**Ans.** 17.45 atm]

8. Calculate the C_{rms} of helium molecules at 300 K.

[**Ans.** 30.82 m s^{-1}]

9. van der Waals constant b for a gas is 4.2×10^{-5} m^3 mol^{-1}. How close can the nuclei of its two molecules come together?

[**Ans.** 1.886×10^{-10} m]

10. Calculate the number of collisions for *HI* molecules at 700 K and 1 atm pressure. Given that $\rho_{HI} = 4 \times 10^{-10}$ m.

[**Ans.** $1.33 \times 10^{28} s^{-1}$ 5^1 m^{-3}]

11. For oxygen, $P_c = 49.7$ atm ; $V_c = 0.0744$ l mol^{-1} and $T_c = 154.4$ K. Find out the pressure of 1 mol of O_2 occupying a volume of 0.27 l at 193 K.

12. For N_2 gas at 1 atm pressure and 298 K, calculate the total number of collisions per second in 1 m^3.

[**Ans.** 8.93×10^{34} s^{-1} m^{-3}]

13. Calculate the root mean square velocity of a gas having a density of 1.25×10^{-3} kg at 352 K and 1 atm pressure.

[**Ans.** 4.9×10^2 ms^{-1}]

Subjective Questions

1. Discuss how the real gases deviate from ideal behaviour.

2. Derive van der Waals equation.

3. Derive a relation between critical constants and van der Waals constants.

4. What is critical phenomena? Give values of critical constants.

5. Write in detail about the liquefaction of gases.

6. Calculate the reduced pressure, volume and temperature of one mole of a gas X which is contained in a 5 l flask at 397 K.

Critical constants of the gas are $T_c = 190.0$ K, $V_c = 0.099$ mol^{-1} and $P_c = 45.5$ atm

[**Ans.** $\pi = 0.1099$, $\theta = 2.09$, $\theta = 50.5$]

7. Calculate the pressure exerted by 3 mol of SO_2 gas contained in a 10 l vessel at 337°C assuming it to be a real gas. van der Waals constants are given as $a = 6.7$ atm l^2 mol^{-2} and $b = 0.056$ l mol^{-1})

[**Ans.** $P = 15$ atm]

8. The critical constants for a gas (mol. wt. = 71) are $P_c = 7.6 \times 10^6$ N m^{-2} and T_c = 417 K. Calculate van der Waals constants for it.

[**Ans.** $a = 0.667$ N m^4 mol^{-2}, $b = 5.7 \times 10^{-5}$ m^3 mol^{-1}]

9. One mole of a gas occupies a volume of 3.8×10^{-4} m^3 at 313 K. Calculate the pressure that the gas would exert assuming it to be (*i*) ideal gas; and (*ii*) real gas.

van der Waals constants are given as

$$a = 0.36 \text{ N m}^4 \text{ mol}^{-2};\ b = 4.28 \times 10^{-5} \text{ m}^3 \text{ mol}^{-1}$$

[**Ans.** (*i*) 6.8×10^6 N m^{-2}; (*ii*) 5.2×10^6 N m^{-2}]

10. Calculate root mean square speed, average speed and the most probable speed of nitrogen gas at 298 K.

[**Ans.** $C_{rms} = 515.2$ ms^{-1}, $c = 473.9$ ms^{-1} $C_{mp} = 422.5$ ms^{-1}]

11. What will be the translation kinetic energy of 16 g oxygen at 273 K?

[**Ans.** 1.7×10^{10} ergs]

12. Calculate the pressure exerted by oxygen gas whose 10^{22} molecules are contained in a vessel of one litre capacity. The root mean square speed of molecules is 4×10^2 ms^{-1}.

[**Ans.** 2.826×10^{-2} Pa]

❑❑❑

2

Liquid State

2.1 INTRODUCTION

It has already been observed that the main characteristics of the gaseous state are random translational motion of molecules and very little intermolecular attraction at normal pressure or temperature values. But in the solids and particularly so in the crystalline state, the individual atoms or molecules are fixed in an ordered arrangement, extending to a long range. However, in liquids, we neither find complete random motion of molecules nor ordered state of arrangement. In liquids the well-ordered arrangement extends over a short-range only, say up to five or six molecular diameter as supported by X-ray studies.

2.2 INTERMOLECULAR FORCES

We have already seen in the gaseous state that attractive forces exist between molecules and liquefication of gases is caused when at high pressure or low temperature molecules are brought very close to each other and are held together to form the liquid.

Various types of interactive forces operating between the liquid molecules are :

1. **Dipole–Dipole Forces**

   These type of intermolecular forces operate between such molecules of liquids which contain permanent dipole moments Fig. 2.1.

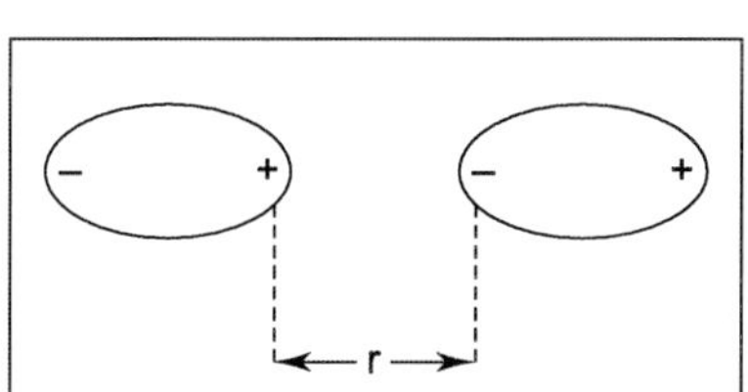

Fig. 2.1 Dipole–dipole forces.

We can now imagine that when two liquid molecules with permanent dipoles come very close to each other the positive end of one molecule gets attracted to the negative end of the other molecule and so on. This force of attraction is electrostatic in nature. Such molecules get aligned due to these dipole–dipole interactive forces and this causes an increase in the boiling point of such liquids.

2. **Ion–Dipole Forces**

When an ion (cation A^+ or anion B^-) is present in the vicinity of liquid molecules with permanent dipole, the cation with +ve charge will align with the negative end of liquid molecules and similarly the anion with –ve charge will align with the positive end of liquid molecules Fig. 2.2.

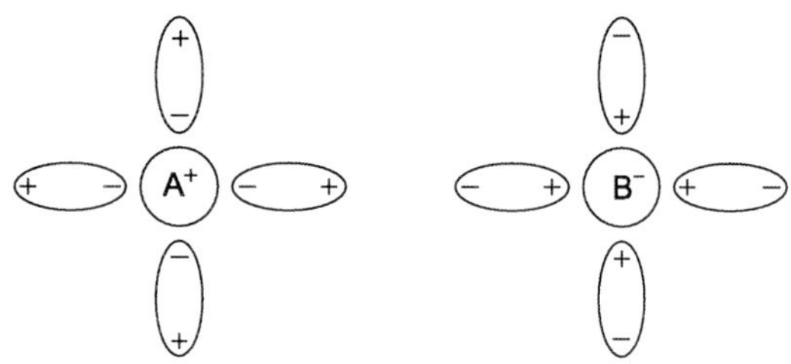

Fig. 2.2 Ion–dipole interaction.

3. **Dipole–Induced Dipole Forces**

When a molecule with permanent dipole comes near to a neutral molecule the electron cloud of the neutral molecule gets distorted in the presence of permanent dipole. This distortion of electron cloud of a neutral molecule is expressed in terms of polarizability as shown in Fig. 2.3.

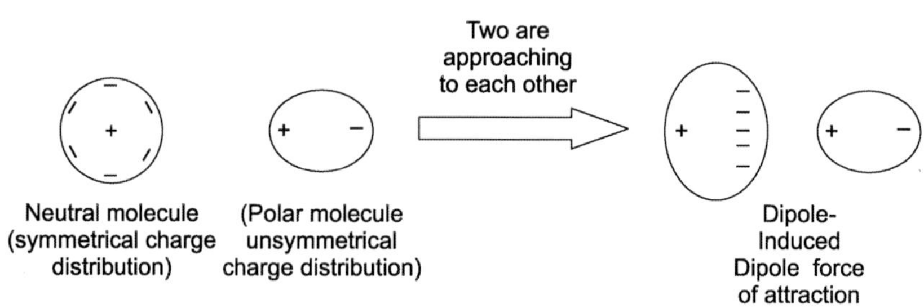

Fig. 2.3 Dipole–Induced dipole interaction.

Thus a polar molecule has induced a dipole in a neutral molecule and these will get attracted by dipole–induced dipole forces.

4. **Induced Dipole–Induced Dipole Attraction – London Forces**

In molecules of rare gases like helium, argon and other permanent gases such as hydrogen and nitrogen, which have been liquefied, it is assumed that in those molecules electron cloud at any instance gets slightly displaced relative to positive charge centre and thus an instantaneous dipole is developed. This instantaneous induced dipole can induce a dipole in the nearby neutral molecule or atom. Such forces of attraction are termed "**London**" **forces**. These are generally short–range forces and their interaction energy–"**London**" **energy**, varies inversely as the sixth power of distance (r) between atoms or molecules.

$$\text{London Energy} = \left(-\frac{k}{r^6}\right) \quad (2.1)$$

where k is a constant.

5. Hydrogen Bonding

It could be observed that boiling point and melting point of hydrides of V, VI and VII group elements (NH_3, H_2O and HF) show an abnormal behaviour indicating that some attractive forces (in addition to van der Waals type) is active amongst them. It has been concluded that hydrogen atom in such molecules is present in between two highly electronegative atoms, when such molecules approach each other, particularly in liquid state. The hydrogen atoms form a hydrogen bond between two oxygen atoms, in water molecules or fluorine molecules, in the case of liquid hydrogen fluoride Fig. 2.4.

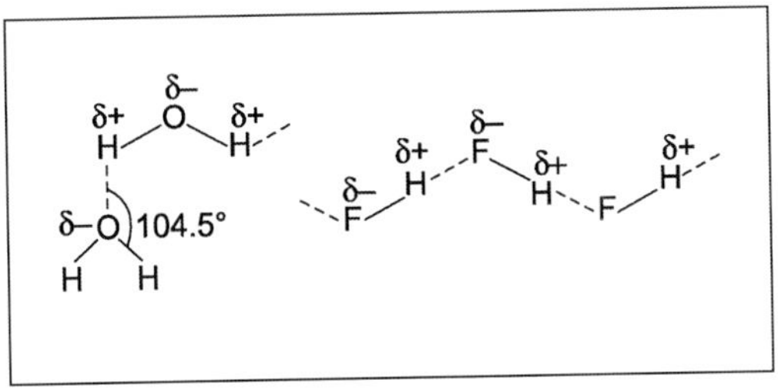

Fig. 2.4 Hydrogen bonding in H_2O and HF.

Thus, two or more molecules may associate together to form a large cluster of molecules $(H_2O)_n$.

Alcohols and carboxylic acids also form associated molecules due to presence of intermolecular or intramolecular hydrogen bonding Fig. 2.5.

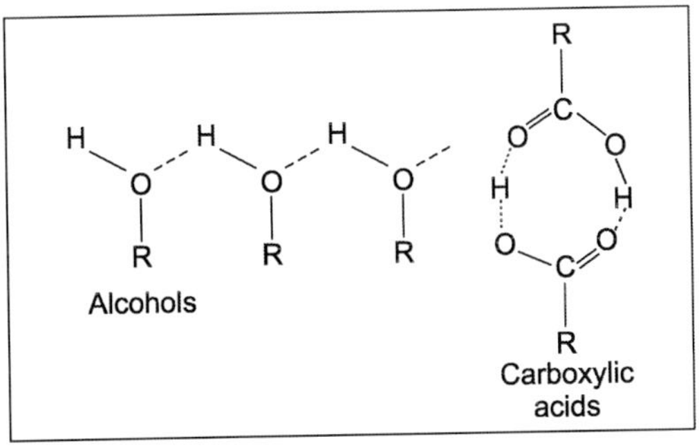

Fig. 2.5 Hydrogen bonding in alcohols and carboxylic acids.

These hydrogen bonds (strength lying between 10 and 40 kJ mol^{-1}) are quite weak in comparison to a covalent bond strength of the order of 400 kJ mol^{-1}.

2.3 STRUCTURE OF LIQUIDS

There are basically three different views developed to explain the structure of liquids:

(*a*) **The Liquids as Modified Gas:** In a stationary liquid, the pressure is same in all directions. It is in this respect we have reason to say that liquids resemble a gas. As we know it is true particularly near critical point when no distinction between liquid and gas exists and it is possible to describe liquid as an imperfect gas.

(*b*) **The Liquid as a Modified Solid:** When solids melt its volume expansion takes place by 10%. This may be either due to an increase in gaps in the packing of the atoms or increase in co-ordination number, *i.e.*, the number of nearest neighbours. However, X-ray studies have revealed that mean intermolecular separation is not changed as much as is expected due to increase in volume. Thus, a liquid closely resembles with the structure of crystalline solids except for creation of certain holes. Liquids are like imperfect solids and holes in liquids correspond to vacant sites of imperfect crystals.

(*c*) **Eyring's Free Volume Theory of Liquids:** Henry Eyring and T. Ree predicted the presence of free volume being occupied by liquid molecules. This free volume is randomly distributed through the liquid and its proportion increases with the rise in temperature Fig. 2.6. The liquid molecules present adjacent to free volume behave like the gas and other liquid molecules far away from free volume possess solid like properties. Thus, properties of liquid depend on the temperature range in which it is studied and behave like a gas–solid combination.

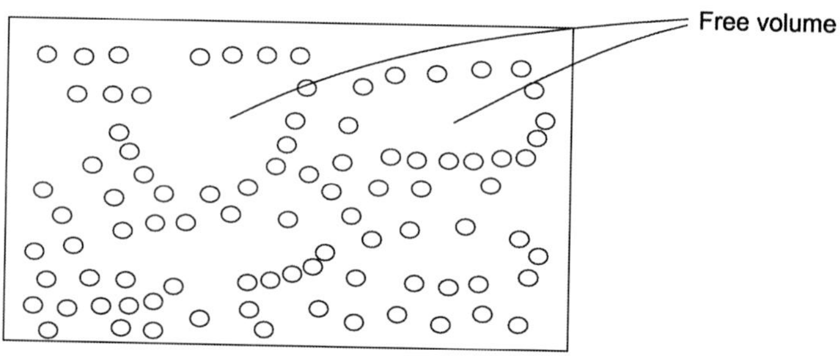

Fig. 2.6 Free volume in liquid state.

(*d*) **Random Packing Model:** Bernel (England) and Scott (Canada) considered the liquid state as a random pack of molecules, recognizing the disordered but relatively close-packed nature of the liquid–structure. Based on these structural features the liquid cannot be expected to have a long–range molecular orders Fig. 2.7.

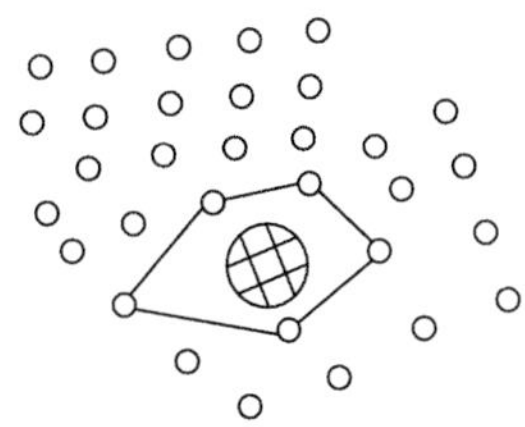

Fig. 2.7 Close pack structure of liquid.

2.4 DIFFERENCE BETWEEN SOLIDS, LIQUIDS AND GASES

When a beam of X-rays is passed through a crystal, characteristic Laue spots appear in diffraction pattern. While in liquids it consists of series of light and dark rings and no coherent diffraction pattern appears in gases. This gives an idea that liquids have some ordered structural features.

A similar conclusion can be presented in a different way by means of radial distribution function (RDF). The distribution function is obtained by selecting a reference molecule/atom within the liquid and drawing a series of concentric spheres around it in a manner such that the volume interval between two neighbouring spheres remain unaltered Fig. 2.8. Thus, we can arrive at the average number of molecules/atom (ρ) per centimetre at a distance r. A plot of radial distribution function ($4\pi r^2\rho$) as a function of r is shown in Fig. 2.9.

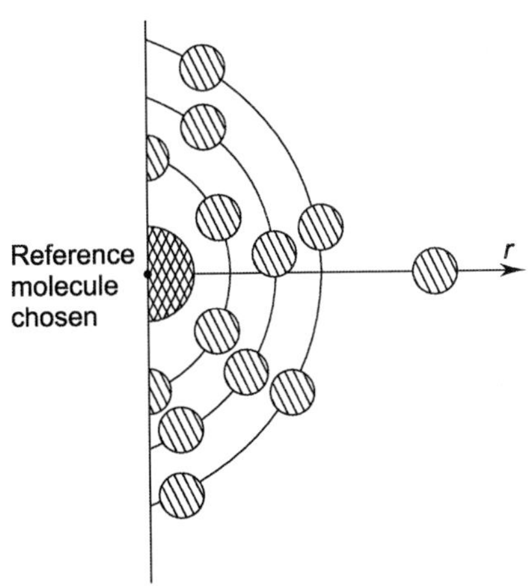

Fig. 2.8 Distribution of molecules in reference to a chosen molecule in the liquid.

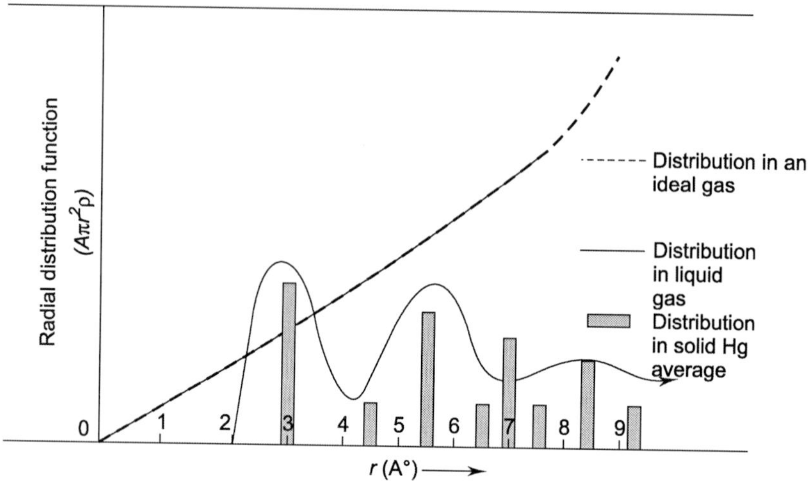

Fig. 2.9 Radial distribution function plotted against radius measured from an arbitrary chosen atom as the origin.

It is observed that with liquid Hg the structure is lost beyond 10 Å while with solid Hg it persists much beyond this value showing existence of long–range orders. though Liquids possess an ordered structure of constituent molecules/atoms, this pattern is very short existing.

2.5 LIQUID CRYSTALS

Liquid crystals are a phase of matter whose order is intermediate (mesomorphic state) between that of liquid and that of a crystal Fig. 2.10. They may be anisotropic or isotropic. This fourth state of matter generally possess orientational or weak positional order and thus reveals physical properties of crystals but flow like liquids. If transitions between the phases are given by temperature, they are called thermotropic. In blends of different components phase transitions may also depend on concentration, and these liquid crystals are called lyotropic. While thermotropics are mostly used for technical applications, lyotropics are important for biological systems, *e.g.*, membranes. Let us take the example of solid phase *p*-cholesteryl benzoate, which on heating fuses sharply to a turbid liquid crystal phase at 145°C, known as its transition point, and then further heating yields clear liquid phase at 178°C, called its melting point. The change may be represented as :

$$\underset{\text{solid phase}}{(p\text{-cholesteryl benzoate})} \xrightleftharpoons{145^\circ C} \underset{\text{liquid phase}}{(p\text{-cholesteryl benzoate})} \overset{178^\circ C}{\rightleftharpoons} \underset{\text{liquid phase}}{(p\text{-cholesteryl benzoate})}$$

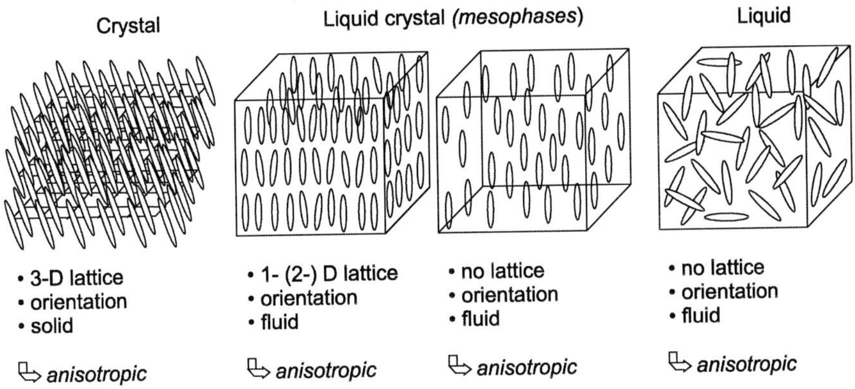

Fig. 2.10 Liquid crystalline mesophases between the solid and isotropic liquid phase.

Chemical structure for two of the most prominent representatives exhibiting liquid crystalline order are given below Fig. 2.11. MBBA is the room temperature liquid crystal and 5 CB, a cyanobiphenyl is the most applicable family of liquid crystals.

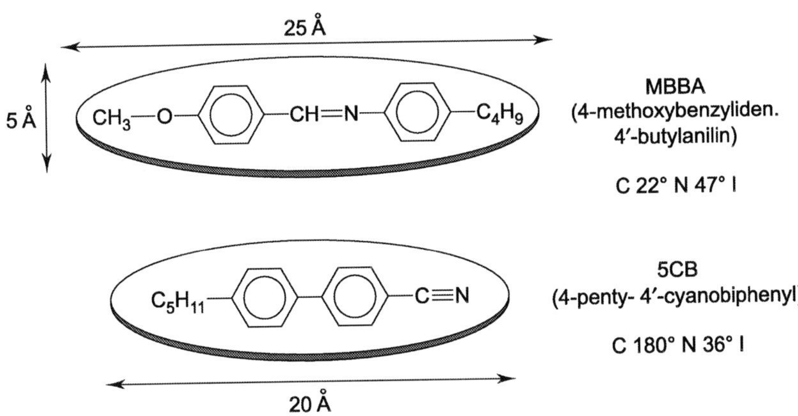

Fig. 2.11 Typical building blocks of liquid crystals.

2.6 DIFFERENCE BETWEEN LIQUID CRYSTAL, SOLID AND LIQUID

As a result of orientational order, most physical properties of liquid crystals are anisotropic, *e.g.*, heat diffusion, magnetic susceptibility, dielectric permittivity or optical birefringence. Additionally, there are two physical qualities, which do not appear in simple liquids, *e.g.*, elastic or frictional torques (rotational viscosity) acting on static or dynamic director deformation, respectively.

2.6.1 Phases of Matter Namely Liquid Crystal, Solid and Liquid

A solid showing the mesomorphic change underges following transformations :

$$\text{solid} \rightleftharpoons \text{liquid crystal} \rightleftharpoons \text{liquid} \rightleftharpoons \text{vapour}$$

Such transformation on vapour pressure—temperature diagram is shown in Fig. 2.12. Vapour pressure curve of solid (crystalline phase) and liquid crystalline phase are distinctly represented by *PR* and *RS* respectively while *SU* is that of isotropic liquid phase. Here we have two triple points, namely *R* and *S* compared to solid, liquid and vapour equilibria phase diagram, where we get only one triple point. At *R* three phases solid crystals, liquid crystals and vapour phase are co-existing, whereas at *S* liquid crystal, liquid and vapour phases are co-existing in equilibrium.

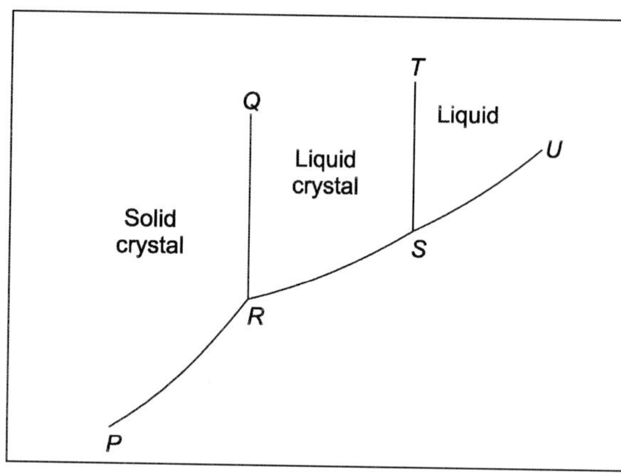

Fig. 2.12 Phase diagram of mesomorphic state.

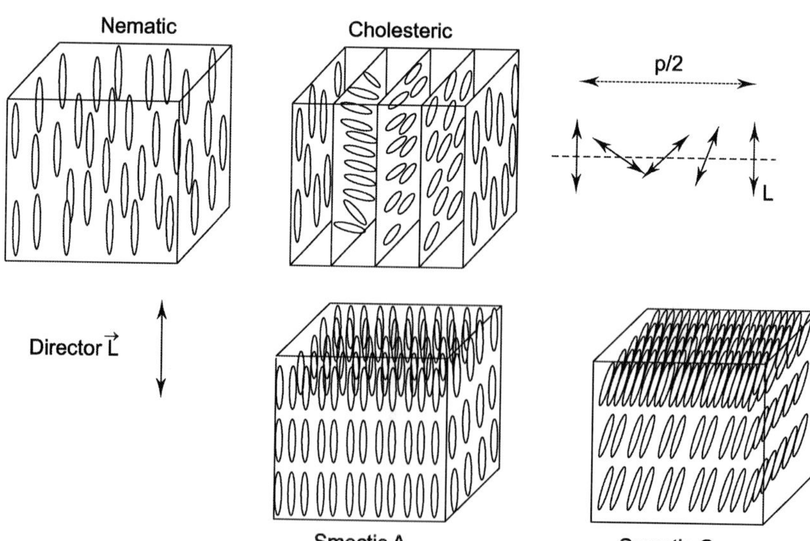

Fig. 2.13 Liquid crystalline mesophases.

2.7 CLASSIFICATION OF LIQUID CRYSTALS

The three liquid crystal phases are shown in Fig. 2.13. In the simplest case the molecules possess only orientational but no positional long–range order. The liquid crystals of this type are called **nematic**, *i.e.*, thread–like texures observed under polarizing microscope. The direction of preferred alignment can be described by a unit vector, the so called nematic director *L* or symmetry axis of the rotational distribution, nematic liquid crystals are uniaxial. The molecular orientations (and hence the material's optical properties) can be controlled with applied electric fields.

If a nematic liquid crystal is made of chiral molecules, *i.e.*, the molecules differ from their mirror image, a **cholesteric** liquid crystal is obtained. This phase exhibits twisting of molecules perpendicular to the director, with the molecular axis parallel to the director. The finite twist angle between adjacent molecules is due to their asymmetric packaging which results in longer-range chiral order.

In the **smectic** chiral phase, the molecules are tilted by a finite angle with respect to the normal layer. The chirality induces a finite azimuthal twist from one layer to the next, producing a spiral twisting of the molecular axis along the layer normal.

The chiral pitch refers to a distance over which the mesogens undergo a full 360° twist. The pitch changes with the alteration of temperature. In some liquid crystal system, the pitch is of the same order as the wavelength of visible light. This causes these systems to exhibit unique optical properties, such as selective reflection and these properties are exploited in a number of optical applications.

In the **discotic phase,** the disk-shaped mesogens can orient themselves in a layer–like fashion known as discotic nematic phase.

2.8 STRUCTURE OF NEMATIC AND CHOLESTERIC PHASES

In nematic phase of liquid crystal, the molecules spontaneously order with long axes roughly parallel. It is characterized by a high degree of long–range orientational order but no translational order. An uniformly aligned nematic has a preferred direction, often described in terms of a unit vector called the director. In cholesteric phase, the distance over which the director rotates to equal and 360°, *i.e.*, chiral pitch, is normally of the order of few hundred nanometres or precisely the wavelength of visible light. This allows the liquid crystals to selectively reflect light of wavelength equal to pitch length, so that a colour will be reflected when the pitch is equal to corresponding wavelength of light in visible spectrum. Changes in the director orientation between successive layers modifies the pitch length resulting in an alteration of the wavelength of reflected light according to the temperature. The angle at which the director changes can be made longer, by increasing the temperature of the molecule and thus it defined the pitch.

2.9 THERMOGRAPHY

Liquid crystal thermography correlates the colour response of a hot surface treated with thermochromic liquid crystal that changes their reflected colour as a function of temperature when illuminated by white light. Two colour temperatures are used for specifying a given liquid crystal, *e.g.*, R40C5W, implies, activation (red colour) temperature at 40°C, a 5 W implies start of blue at 5°C above red. The liquid crystal material is encapsulated in a 5–10 micron sphere suspended in a water-based binder material, which provides good protection. To measure with a liquid crystal, the relationship between temperature and colour response must be calibrated. A temperature of 39.9°C is recognized by a system with Red (R), Green (G) and Blue (B) values as 107.9, 115.7 and 93.1. This system can now translate the RGB values into actual temperature. The specimen surface is cleaned with alcohol. A thin and uniform coat of black paint is applied to the test specimen and calibration surface. It is then dried and desired thermochromic liquid crystal is sprayed to both surfaces. A white light, free from IR and UV radiation, is required to obtain reliable reflected light intensity from thermochromic liquid crystal coated surface Fig. 2.14.

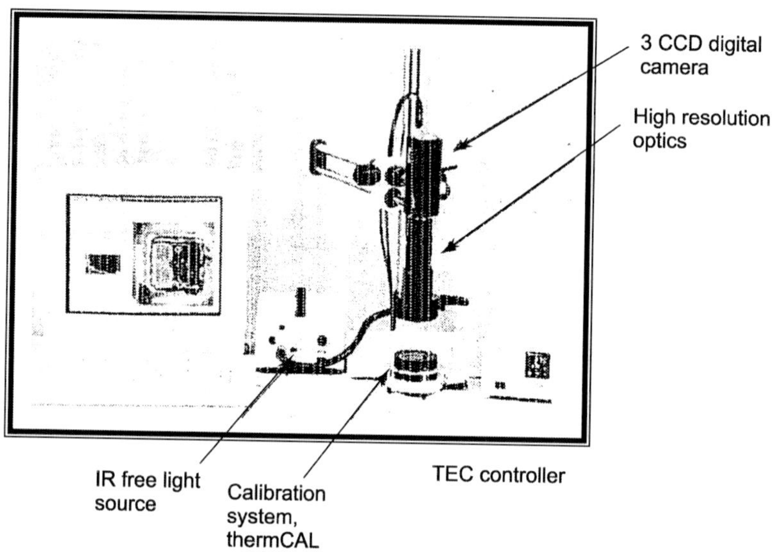

Fig. 2.14 Liquid crystal thermography.

2.10 SEVEN–SEGMENT CELL

One common requirement for many different digital devices is a visual numeric display. A matrix of 28 LEDs (light emitting diodes) or LCDs (liquid display units) in a 7×4 array can be used. We can then light up selected LEDs or LCDs in the pattern required for whatever character we want. A much better way is to arrange the minimum possible number of LEDs / LCDs in such a way as to represent only numbers. This requires just seven LEDs (plus an eighth one for the decimal point). A common technique is to use a shaped piece of translucent plastic to operate as a specialized optical fibre, to distribute the light from LED/LCD evenly over a fixed bar shape. The seven bars are laid out as a squared-off figure "8". The result is known as a seven segment LED/LCD.

2.10.1 Seven-Segment Display Layout

Basic layout of the seven segments in a seven-segment display is given in Fig. 2.15. The seven-segment digits also include a decimal point ("*dp*") and some also include an extra triangle to turn the decimal point into a comma. This improves readability of large numbers on a calculator, for example, the decimal point is shown on the right. In addition most displays are actually slanted a bit, making them look as if they were in italics. This arrangement allow us to turn one digit upside down and place it next to another, so that the two decimal point looks like a colon between two digits. Seven-segment displays can be packaged in a number of ways. A typical 14-pin DIP package to be installed vertically is shown in Fig. 2.15.

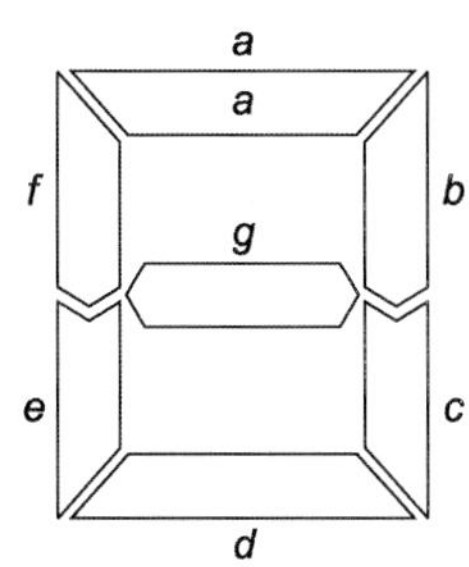

Fig. 2.15 Seven-segment display.

Seven-segment display can be constructed using (*a*) fluorescent displays (in microwave oven), (*b*) Liquid crystal display (LCDs), (used in many battery powered devices such as watches) and (*c*) LCDs.

2.10.2 Schematic Diagram

Either all the cathodes or all the anodes are connected together into a common lead, while the other end of each LED/LCD is individually available (Fig. 2.16). This means fewer electrical connections to the package, also allows us to easily enable or disable a particular digit by controlling the common lead in liquid crystal display (LCD). A thin flat display device is made up of any number of colour or monochromic pixels arranged in front of a light source or reflector. Each pixel of an LCD consists of a layer of molecules aligned between two transparent electrodes and two polarizing filters, axis of transmission of which are perpendicular to each other. With no liquid crystal between the polarizing filters, light passing through first filter would be blocked by the second (crossed) polarizer. Electrodes are made of a transparent conductor, indium tin oxide (ITO). When the voltage is applied across the electrodes, a torque acts to align the liquid crystal molecules parallel to the electric field, distorting the helical structure. This reduces the rotation of the polarization of the incident light, and the device appears grey.

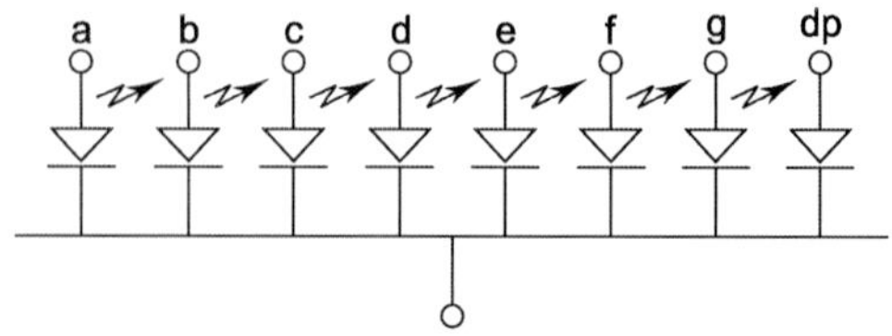

Fig. 2.16 Schematic diagram of LCDs.

Objective Type Questions

1. Density of a substance in the liquid state is

(*a*) lower than in gaseous state (*b*) much higher than in gaseous state

(*c*) same as in the solid state (*d*) none of the above.

2. Considerable dipole-dipole forces in liquid will

(*a*) increase its boiling point (*b*) lower its boiling point

(*c*) produce liquid crystal (*d*) change the mean free path of molecules.

3. The free volume in liquid is the volume in which liquid molecules

(*a*) can move freely (*b*) form liquid crystals

(*c*) move either to solid state (*d*) cannot move.

4. A liquid crystal has properties

(*a*) more like a solid crystal than liquid

(*b*) more like gas

(*c*) more like liquid than solid crystal

(*d*) none of these.

5. Fill in the blank in the following sentence : "A __________ has ordered pattern within a short distance from any given molecule".

(*a*) gas (*b*) liquid

(*c*) solid (*d*) none of these.

6. The change shown here is

$$p\text{-cholesteryl benzoate (solid)} \underset{}{\overset{145°C}{\rightleftharpoons}} p\text{-cholesteryl benzoate (liquid)}$$

(*a*) melting point (*b*) boiling point

(*c*) transition point (*d*) none of these.

7. Smectic transformation is shown by

(*a*) ethyl-*p*-azoxy benzoate (*b*) *p*-azoxy phenetole

(*c*) *p*-cholesteryl benzoate (*d*) salicylaldehyde.

8. *p*-Cholesteryl benzoate is a liquid crystal of type
(*a*) nematic (*b*) cholesteric
(*c*) smectic (*d*) none of these.

9. Radial probability distribution of molecules in liquid state show
(*a*) parabolic radial distribution curve
(*b*) vertical lines
(*c*) maxima and minima in distribution curve
(*d*) all of these.

10. When X-rays are passed through a liquid, we observe
(*a*) presence of characteristic laue spots
(*b*) series of light and dark rings
(*c*) no coherent diffraction
(*d*) all of these.

ANSWERS

1. (*b*)	**2.** (*a*)	**3.** (*a*)	**4.** (*a*)	**5.** (*b*)
6. (*c*)	**7.** (*a*)	**8.** (*a*)	**9.** (*c*)	**10.** (*a*)

Short Answer Type Questions

1. Define liquid state.
2. What is a condensed state of matter ?
3. Explain hydrogen bonding.
4. Discuss radial distribution function.
5. Explain random packing model.
6. What is anisotropy ?
7. Explain isotropic behaviour.
8. Define transition point.
9. Why soap when present in water appears turbid ?
10. What is mesomorphic state ?

Subjective Questions

1. Discuss the various types of intermolecular forces in liquid.

2. What is meant by radial distribution function of liquids ? How the structure of liquids can be explained using radial distribution function ?

3. What are structural features of nematic and cholesteric liquid crystals ?

4. How are the liquid crystals classified?

5. Give structural differences between solid, liquid and liquid crystals.

6. Write short notes on:

(*a*) Ion-dipole forces (*b*) Dipole-induced dipole forces

(*c*) Induced dipole-induced dipole forces.

7. Explain in brief Eyring free volume theory of liquids.

8. What are nematic liquid crystals ? When can a nematic liquid crystal may be classified as cholesteric liquid crystal ?

9. Write an essay on thermography.

10. Discuss the display layout of a seven segment cell having LEDs.

❑❑❑

3

Solid State

3.1 INTRODUCTION

Solids are characterized by (*i*) definite shape ; and (*ii*) definite volume.

As a consequence solids are almost incompressible. Depending on the regularity of arrangement of atoms or molecules, solids may be classified as (*a*) **amorphous,** *e.g.*, glass, pitch, rubber, etc., where the arrangement is rather irregular and (*b*) **crystalline**, *e.g.*, sugar, common salt, etc., where the molecules or atoms are in a regular arrangement and as a result solids aquire a crystalline nature.

Nowadays the term solid is used for solid substances which are crystalline in nature. A crystalline solid is characterized by (*a*) definite melting point; (*b*) definite crystal lattice ; and (*c*) definite enthalpy of formation. A crystal is a three–dimensional arrangement of the constituent unit which may be atoms, molecules or ions.

3.2 TYPES OF CRYSTALLINE SOLIDS

Crystalline solids are classified into following four groups depending on the nature of forces that hold the particles together.

(*i*) **Metallic Solids :** In metallic solids the positive ions of the atoms are held together by a surrounding sea of electrons. The force of attraction known as metallic bonding between the positive ions of the metal and electrons is responsible for cohesion in the metal.

Metallic bonding occurs in metals, *e.g.*, Cu, Na, Fe, etc., and explains unique properties of metals like thermal and electrical conductivity, malleability and metallic lusture.

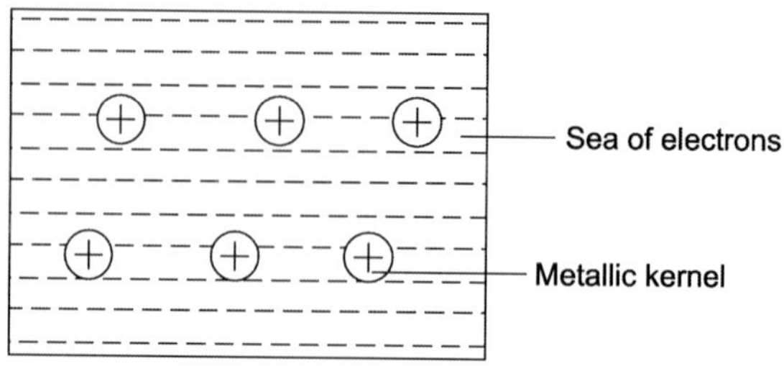

Fig. 3.1 Structure of metallic solids.

(ii) **Ionic Solids:** Ionic solids like NaCl, CaO *etc.*, are made up of positive and negative ions which are held together by electrostatic attraction between oppositely charged ions and is known as ionic bond.

Electrostatic attraction between ions of opposite charges results in the efficient packing of two sets of ions (or spheres) of different radius or sizes. The geometric requirements for close packing of spheres of different sizes is given by $\frac{r_a}{r_c}$, where r_c is the radius of smaller ion (usually cation) and r_a the radius of bigger ion (usually anion).

Ionic compounds are brittle and have high melting points.

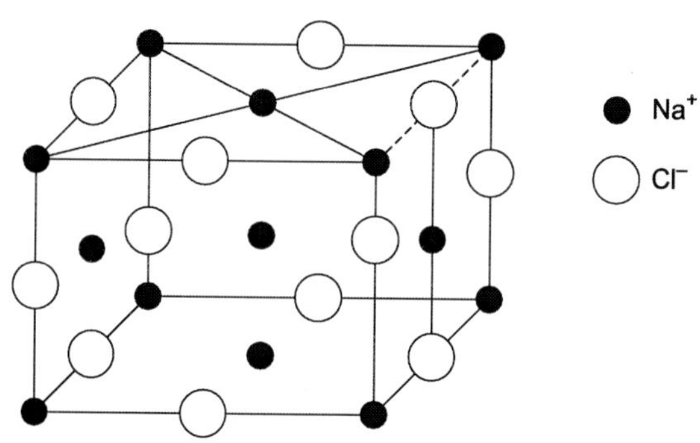

Fig. 3.2 Sodium chloride (ionic crystal).

(iii) **Molecular Solids:** In molecular solids the molecules or atoms are held together by weak intermolecular forces of attraction. These forces originate from dipole–dipole induction and London dispersion and are collectively called van der Waals forces.

Molecular solids are soft, volatile and are exemplified by solid neon, solid carbon dioxide, ice, etc.

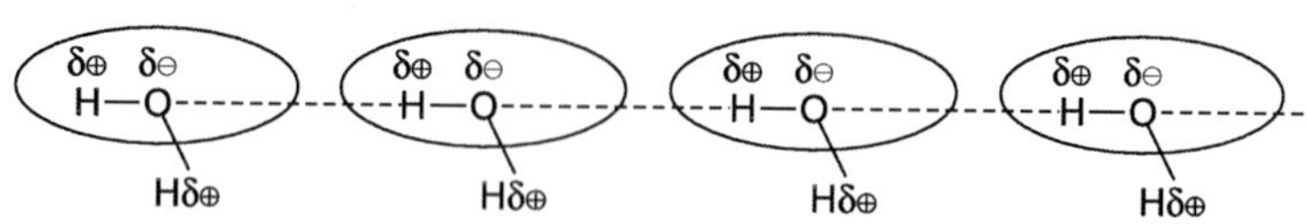

Fig. 3.3 Molecular solids with dipole–dipole forces of attraction.

(*iv*) **Covalent Solids:** Covalent solids consist of atoms held together by covalent bonds. Usually this results in a large network or chain. Examples of covalent solids are diamond, carborundum (SiC), asbestos, etc.

Covalent solids are hard, non-conducting and have high melting points.

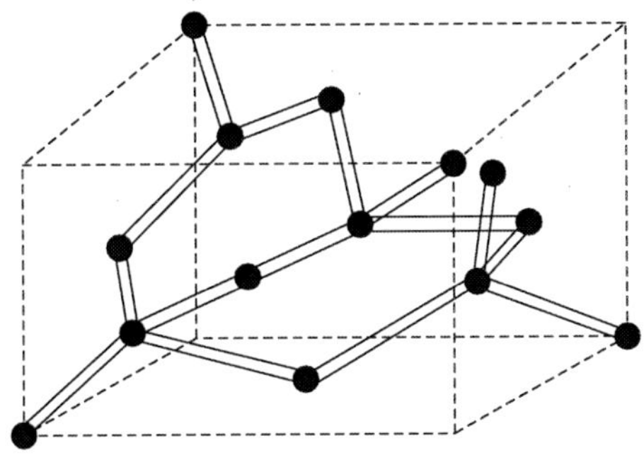

Fig. 3.4 Tetrahedral structure of diamond crystal.

3.3 SPACE LATTICE (OR CRYSTAL LATTICE) AND UNIT CELL

A crystal is a three-dimensional ordered arrangement of constituent units (*i.e.*, atoms, molecules or ions). This ordered arrangement of a crystal is best described in terms of space or crystal lattice.

The space lattice of a crystal may be considered to be made up of three dimensional basic pattern called the **unit cell**. The unit cell is the basic unit from which a crystal may be constructed by stacking these units in three dimensions. Hence, external appearance of a crystal is often determined by the shape and size of the unit cell.

There are only seven basic shapes possible for unit cells which gives rise to seven crystal systems. Many of these crystal systems have more than one possible crystal lattice. Thus constituent particles may be present only at corners of unit cell (simple or primitive lattice) or within the body of the unit cell or on the faces of the unit cell.

Thus each crystal belonging to a given system has unit cells with one of the seven shapes with the constituent particles present at corner, body or face of the cell. [In all 14 different lattices known as **Bravais lattices** have been observed in all known crystals.] These are shown below :

Crystal system	Edge length	Angles	Shape
1. Cubic Examples : Cu, Na, NaCl, CsCl, etc.	$a = b = c$	$\alpha = \beta = \gamma = 90°$	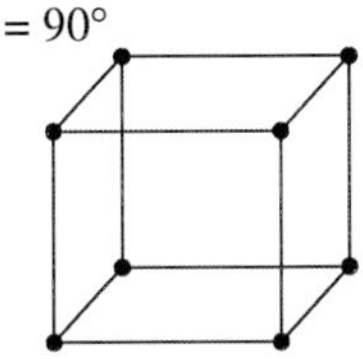 Simple cubic 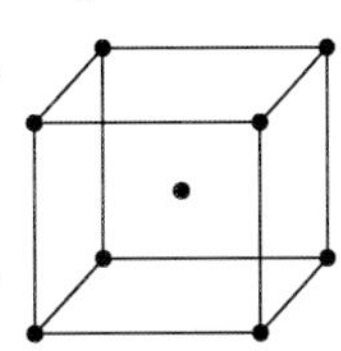 Body-centered cubic

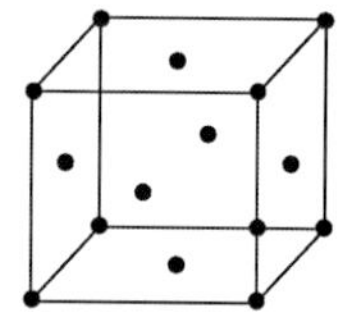

Face–centered cubic

2. Tetragonal $a = b \neq c$ $\alpha = \beta = \gamma = 90°$
   Examples :
   TiO_2 (rutile), Sn

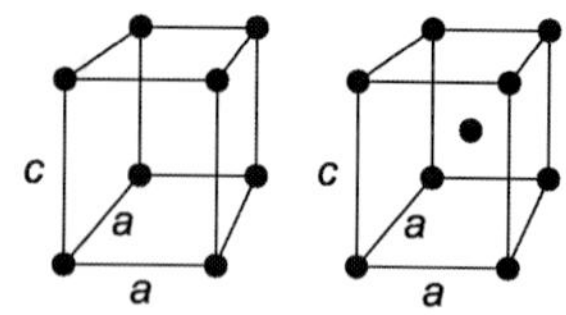

Simple Body-centered

Tetragonal

3. Orthorhombic $a \neq b \neq c$ $\alpha = \beta = \gamma = 90°$
   Examples :
   $BaSO_4$, $CaCO_3$ (argonite)

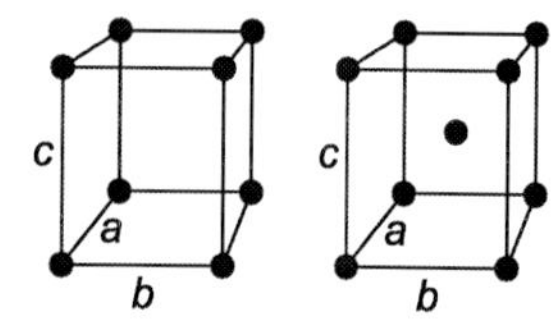

Simple Body-centered

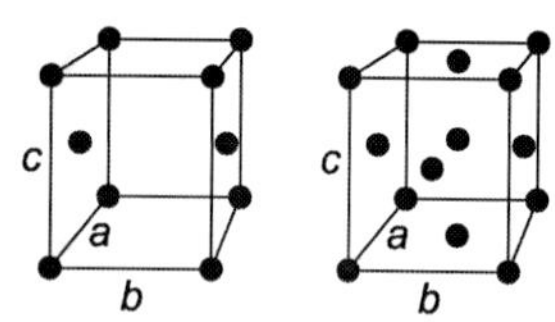

End-centered Face-centered

Orthorhombic

4. Monoclinic $a \neq b \neq c$ $\alpha = \beta = 90°, \gamma \neq 90°$
   Examples :
   S, $Na_2B_4O_7 \cdot 10\ H_2O$ (borax)

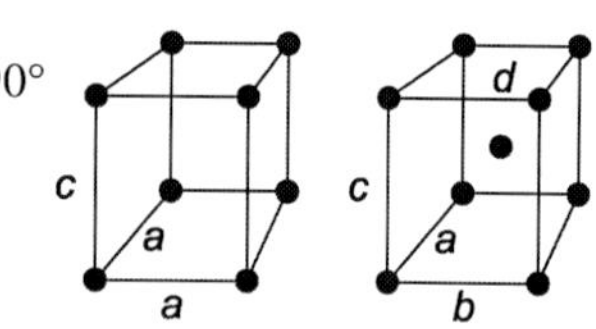

Simple Body-centered

Monocilinic

5. Hexagonal $a = b \neq c$ $\alpha = \beta = 90°, \gamma = 120°$
   Examples :
   Graphite, ZnO

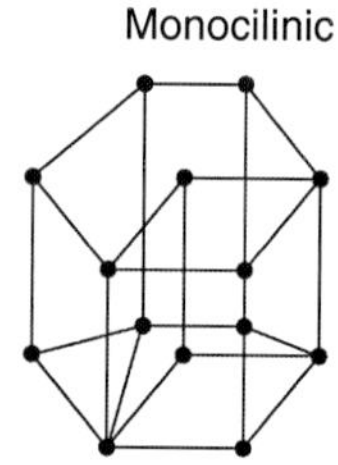

Simple hexagonal

6. Rhombohedral $a = b = c$ $\alpha = \beta = \gamma \neq 90°$

Examples :

HgO, $CaCO_3$ (calcite)

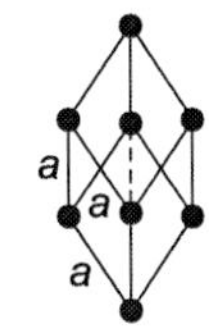

Simple rhombohedral

7. Triclinic $a \neq b \neq c$ $\alpha \neq \beta \neq \gamma \neq 90°$

Examples :

$CuSO_4 \cdot 5H_2O$, $K_2Cr_2O_7$

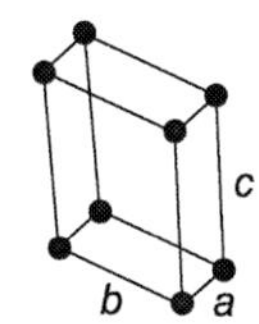

Simple triclinic

3.4 POLYMORPHISM AND ISOMORPHISM

Many substances are known to occur in more than one solid form. These substances undergo a change from one crystalline form to another. The existence of a substance in more than one crystalline form is known as **polymorphism.** Thus carbon exists either as diamond or as graphite. Fluorescein exists in a yellow crystalline form or orange crystalline form.

Polymorphism occurring in elements is also known as **allotropy.** Thus, sulphur exists as rhombic or monoclinic sulphur which are its allotropes.

One form of a polymorph can be converted into another at a definite temperature known as *transition temperature.* If the melting point of one form is above the transition temperature then both forms can exist independently and this phenomenon is known as **enantiotropy.** Thus, sulphur rhombic has a transition temperature of 95.6°C when it changes to sulphur monoclinic while its melting point is 120°C. Thus, both rhombic and monoclinic forms can exist independently.

Isomorphism: When different substances crystallize in the same form, the phenomenon is known as isomorphism and substances are called isomorphs. A common example of isomorphism is alum which have the general formula $M_2SO_4 \cdot R_2(SO_4)_3 \cdot 24H_2O$ where M is monovalent cation and R is trivalent cation exemplified by:

Ferric alum $(NH_4)_2\ SO_4 \cdot Fe_2\ (SO_4)_3 \cdot 24H_2O$

Chrome alum $K_2SO_4 \cdot Cr_2\ (SO_4)_3 \cdot 24H_2O$

Potash alum $K_2SO_4 \cdot Al_2\ (SO_4)_3 \cdot 24H_2O$.

Study of isomorphous compounds led Mitscherlitsch to give Mitscherlitsch law of isomorphism.

"Substances with identical crystal structure must have elements with identical valencies and equal number of atoms united in identical manner".

3.5 NUMBER OF ATOMS/MOLECULES/IONS PER UNIT CELL

A particle at the corner of the cube is shared by eight cubes. Hence its contribution in the cube is $\frac{1}{8}$, but there are eight such positions. So number of particles in the unit cell $= \frac{1}{8} \times 8 = 1$.

A particle in the body of the cube is not shared by any other cube. So its contribution in the unit cell $= 1 \times 1 = 1$.

A particle at the face of the cube is shared by two cubes. So its contribution in the cube is $\frac{1}{2}$ but there are six such faces. Hence, number of particles in the unit cell $= \frac{1}{2} \times 6 = 3$.

3.6 CRYSTALLOGRAPHY

It involves the study of internal structure and physical properties of the crystals. There are three fundamental laws governing the crystal systems :

(*i*) Law of constancy of interfacial angles (**Steno's law**)

(*ii*) Law of rationality of indices (**Hauy's law**)

(*iii*) Law of symmetry.

3.6.1 Law of Constancy of Interfacial Angles

The law states that the angle of intersection of any two corresponding faces is always same for any crystal of the same substance.

3.6.2 Law of Rationality of Indices

The law states that it is possible to choose along three coordinate axes unit distances (represented by a, b, c) such that ratio of these three intercepts of any plane of the crystal is given by $ma : nb : pc$, where m, n and p are either integral whole numbers including infinity or fractions of the whole numbers and are known as **Weiss indices.**

The Weiss indices have now been replaced by **Miller** indices which are obtained by taking the reciprocals of the Weiss indices and then by multiplying throughout by the smallest number to get all the reciprocals as integral numbers and are represented by h, k, l.

Thus, for a plane for which Weiss indices m, n and p are given by $\frac{1}{2}a : b : \alpha c$ becomes $2a : b : 0c$ or simply as $2 : 1 : 0$ in Miller indices (h, k and l). If the plane makes an intercept on negative side, *e.g.*, $-a, -b, -c$ the Miller indices, h, k, l will be $\bar{2} : \bar{1} : 0$, the bar above indicating interaction on negative side.

Spacing of Lattice Planes in a Cubic Lattice: In Fig. 3.6, a unit cubic cell having constituent units at each corner is shown.

Here consider *ABCD* as a (100) plane. There are six such planes corresponding to Miller's indices in this cube corresponding to each face of the cube. These are 100, 010, 001, $\bar{1}00, 0\bar{1}0, 00\bar{1}$ *DCFE* is a (110) plane and there are 12 such planes in the cube. These are 110, 011, 101, $\bar{1}10, 0\bar{1}1, \bar{1}01, 1\bar{1}0, 01\bar{1}, 10\bar{1}, \bar{1}\bar{1}0, 0\bar{1}\bar{1}, \bar{1}0\bar{1}$ *EDH*, is (111) plane and there are eight such planes in the cube. These are 111, $\bar{1}11, 1\bar{1}1, 11\bar{1}, \bar{1}\bar{1}1, \bar{1}1\bar{1}, 1\bar{1}\bar{1}, \bar{1}\bar{1}\bar{1}$.

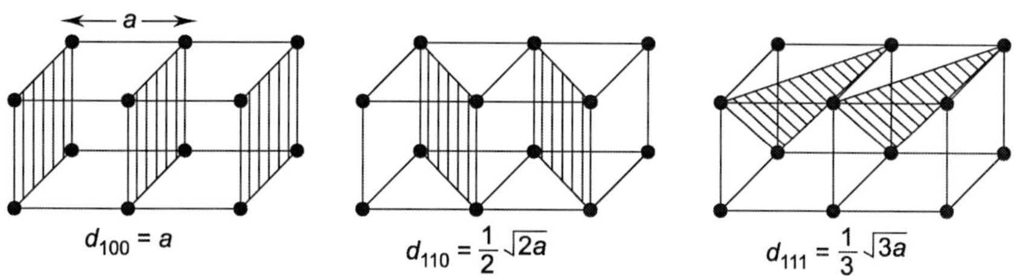

Fig. 3.5 Lattice planes and separations in simple cubic system.

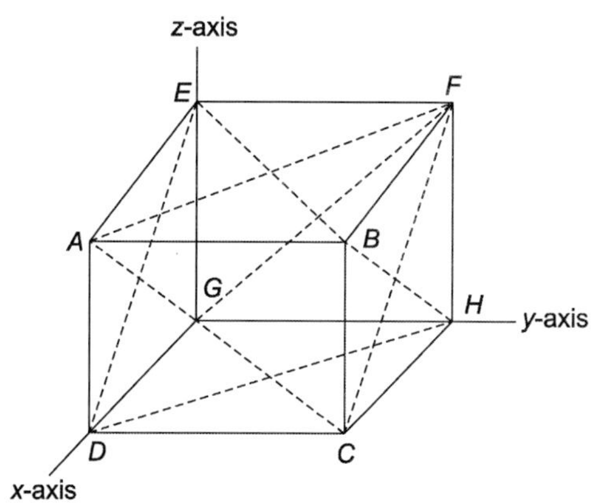

Fig. 3.6 Cubic unit cell.

If one side of the unit cube is considered then this represents the separation of successive (100) planes, as well as of other planes, *e.g.*, *HCBF* (010) plane and *EFCD* (110) plane. The distances between two planes are in the ratio $\sqrt{\frac{1}{h^2+k^2+l^2}}$.

If '*a*' is the edge length of the cube, then it can be shown that distance between successive planes of (110) type is $\frac{a}{\sqrt{2}}$ and the perpendicular separation of *EDH* (111) and *AFC* (111) planes is $\frac{a}{\sqrt{3}}$.

From the above it is clear that the separation of (100), (110) and (111) planes of a cubic lattice is in the ratio $1 : \frac{1}{\sqrt{2}} : \frac{1}{\sqrt{3}}$.

Law of Symmetry: The law states that all crystals of the same substance always have same elements of symmetry. The symmetry elements in a crystal may be with respect to a plane (plane of symmetry), axis (axis of symmetry) or centre (centre of symmetry).

Total number of elements of symmetry (plane + axis + centre of symmetry) is called element of symmetry. A cubic unit cell has 23 elements of symmetry which includes one centre of symmetry, nine planes of symmetry and 13 axes of symmetry (3 four–fold axes, 4 three-fold axes and 6 two-fold axes of symmetry).

Problem 1. Crystal plane intercepts of the three crystallographic axes are in the multiples of the unit distances 2, 3 and 1. Calculate the Miller indices of this plane.

Solution :

Let the unit length be 'a', then intercepts made on x, y and z axes are $2a$, $3a$ and a, respectively.

Thus the ratio of the intercepts are

$$\frac{a}{2a} : \frac{a}{3a} : \frac{a}{a} \quad \text{or} \quad \frac{1}{2} : \frac{1}{3} : 1$$

Hence, the Miller indices (h, k, l) are

$$h : k : l = \left(\frac{1}{2} : \frac{1}{3} : 1\right) \times 6 = 3 : 2 : 6$$

Thus, the Miller indices of plane are (326).

3.7 X-RAY DIFFRACTION BY CRYSTALS AND BRAGG'S LAW

Diffraction of X-rays by crystals was first studied by W.H. Bragg and W.L. Bragg when they observed that it provided structural information regarding the arrangement of constituent particles (atoms, ions or molecules) in the crystal.

The basis of X-ray diffraction crystallograpy is that every constituent particle is able to scatter X-rays to an extent dependent on the number of electrons and planes that have a constituent particle. In other words, every lattice plane behaves towards X-rays in a fashion similar to that of a line in diffraction grating towards visible light. Such that the nature of X-ray diffraction by the crystal is determined by the spacing between successive planes.

Suppose a beam of monochromatic X-ray of wavelength λ is incident on a crystal. After reflection from the planes *AA* and *BB* it goes along *GH* and *FI*, respectively as shown in the figure below.

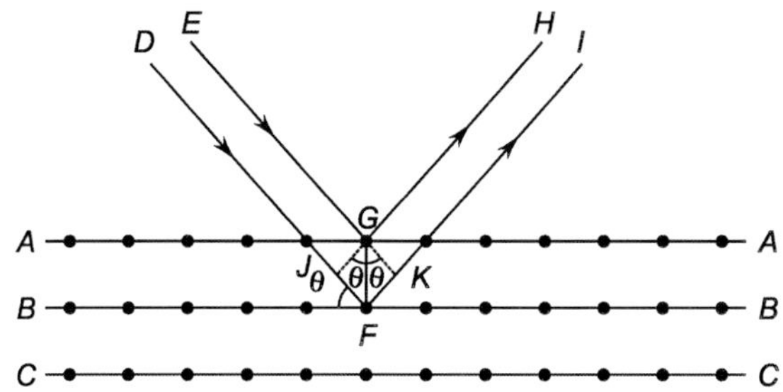

Fig. 3.7 X-ray diffraction by crystal.

The X-rays are reflected according to laws of reflections. However, the intensity of reflected beam will be maximum at certain angle, *i.e.*, when the two reflected waves from two different planes have a phase difference equal to integral multiple of the wavelength of the X-ray.

Let the lattice spacing between two successive planes *AA* and *BB* be '*d*' and the glancing angle of incident X-ray be 'θ'.

Then the path difference between two reflected waves along *GH* and *FI* will be equal to $JF + FK$.

$$\angle GFB = 90° \text{ and } \angle JFB = \theta$$

Hence in ΔGJF

$$\angle JFG = 90° - \theta \quad \text{and} \quad \angle FGJ = \theta$$

Similarly in ΔGFK, $\angle FGK = \theta$

In the ΔGJF,

$$\sin\theta = \frac{JF}{GF} \text{ or } JF = GF \sin\theta$$

but $GF = d$

Hence $JF = d \sin\theta$

Similarly $FK = d \sin\theta.$

$$\text{Path difference} = JK + FK = 2d \sin\theta.$$

When the path difference between the two waves is an integral multiple (n) of wavelength (λ), the interference will reinforce each other and the reflected beam will be of maximum intensity. Hence,

$$n\lambda = 2d \sin\theta$$

where $n = 1, 2, 3, 4.$

The equation is known as **Bragg's equation.**

When $n = 1$, the beam gives the diffraction of the first order and if the wavelength of X-rays which produces maximum intensity at a glancing angle θ is known, then the distance between the lattice planes of crystal, *i.e.*, '*d*' can be calculated.

The X-ray spectrum for two wavelengths λ_1 and λ_2 in the first, second and the third order is shown in Fig. 3.8. It is evident from the figure that

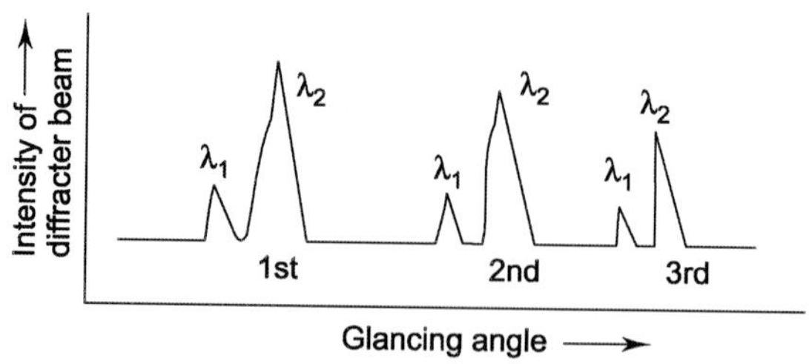

Fig. 3.8 X-ray spectrum.

(*a*) The intensity of a line corresponding to a wavelength decreases with the order of spectrum.

(*b*) The separation between the lines corresponding to two lines of near-wavelength increases with the order of spectrum.

3.8 DETERMINATION OF CRYSTAL STRUCTURE EXPERIMENTALLY

3.8.1 Bragg's Method

In this method a beam of X-ray of definite wavelength (λ) after passing through a slit is incident upon the face of the crystal which is mounted on a rotating table. The rays reflected by the crystal pass through another slit into a detector based on ionization power of radiations.

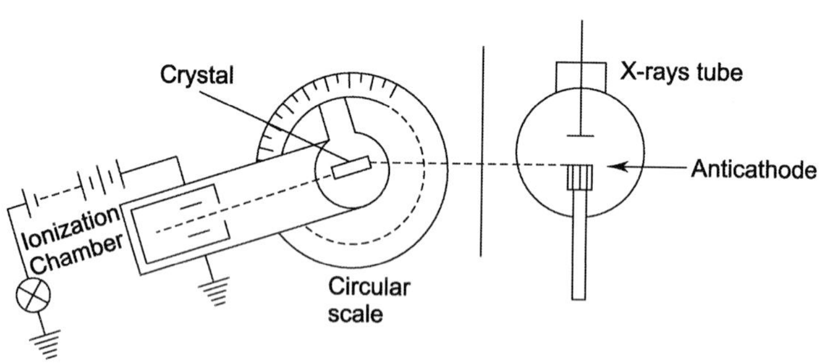

Fig. 3.9 Bragg's spectrometer.

Initially, the glancing angle between the X-rays and the crystal face is kept small and is increased gradually by rotating the table.

The glancing angle θ at which the strongest reflections are observed is noted and is used to calculate '*d*'– the distance between planes.

Drawbacks of the method are: (*i*) a large crystal with well-defined faces is required and (*ii*) reflections of several orders must be noted for several faces – thus consuming lot of time.

3.8.2 Laue's Method

In this method a simple crystal is mounted on a stand with crystal axis perpendicular to the X-ray beam.

A beam of X-rays after passing through a slit-collimator is allowed to fall upon the crystal. After diffraction the X-rays fall on a photographic plate which records the intensities and position of diffracted rays. The constructive reinforced maximum diffractions are noted and Bragg's equation is used to find plane spacings.

The advantage of photographic method is that the symmetry of constituent particles is mostly observable in the photographic plate but the method is not very suitable for finding unknown lattice parameters as it is difficult to identify wavelengths corresponding to specific spots in the photographic plate.

3.8.3 Powder Method

It was given by Debye and Scherrer and it uses powdered crystal rather than a single crystal. The argument is that powdered crystal consists of a multitude of minute crystals.

The crystal powder is taken in a capillary tube or deposited on a fibre and mounted on a rotating table. The sample is gradually rotated in an incident beam of X-rays at all possible orientations. The cone of reflected radiations produced is recorded on a photographic plate as curved lines.

For every set of planes powder specimen reflected beam shows a cone of scattered radiations on the photographic plate as shown in Fig. 3.10 (*a*, *b*).

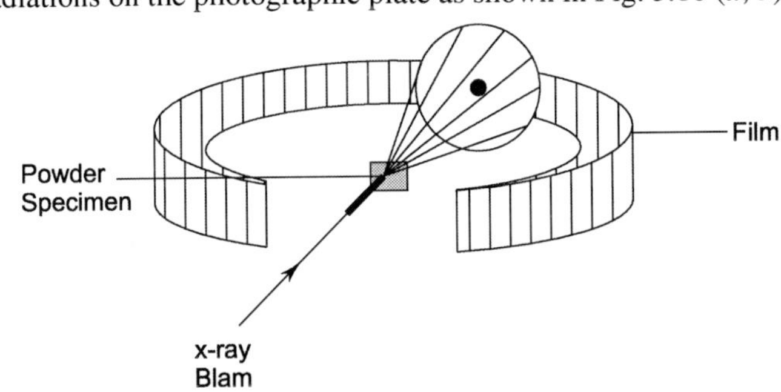

Fig. 3.10(*a*) Powder method for X-ray diffraction.

Fig. 3.10(*b*) A powder photograph.

Using Bragg's equation and noting the distances between lines on the film, the lattice plane spacings can be calculated.

The advantage of powder method is that it does not require a large single crystal and it can also be used to distinguish crystalline substance from amorphous substances.

3.8.4 Rotating Crystal Method

In this method a small crystal is rotated about an axis parallel to one of the crystal axes and a beam of X-rays is incident upon it at right angles.

As the crystal rotates different planes are incident to X-rays and the diffraction produces spots on the photographic plate.

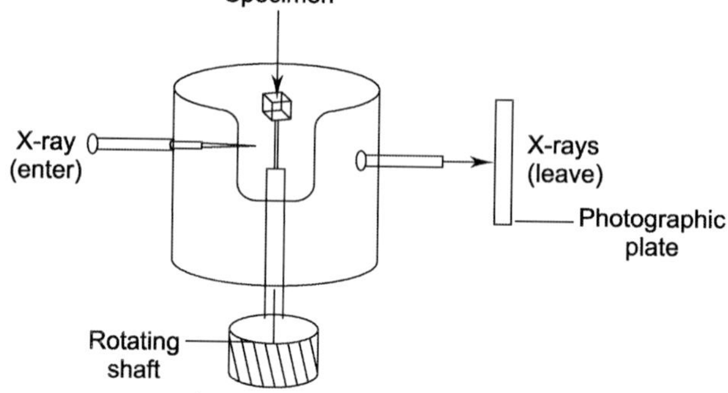

Fig. 3.11 Rotating crystal method.

Photographs are taken with crystal rotating about each of the three axes and using Bragg's equation lattice spacing and the size of unit cell may be derived. The method is most modern, widely applicable and specifically useful for studying complex crystals.

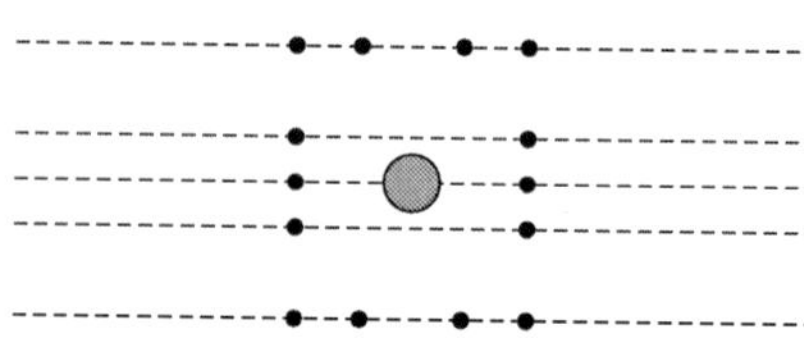

Fig. 3.12 Rotating crystal photograph.

3.9 CALCULATIONS INVOLVING UNIT CELL DIMENSIONS

The structure and dimensions of a unit cell can be determined by X-ray diffraction method described above. Knowing these it is possible to calculate volume, density and mass of atoms in a unit cell using the following formula :

$$\rho = \frac{nM}{NV}$$

where, ρ = theoretical density of the solid

n = number of particles (atoms, ions, molecules) in a unit cell

M = molar mass of particles

N = Avogadro number

V = volume of unit cell

For cubes $V = a^3$, where, a = cube edge length.

Problem 2. The density of Li metal is 0.53 g cm^{-3} and the separation of the (100) planes of the lattice is 350 pm. Determine whether the lattice is *fcc* or bcc (given that atomic weight Li = 6.94 g mol^{-1}).

Solution : Density

$$\rho = 0.53 \text{ g cm}^{-3} \equiv 530 \text{ kg m}^{-3}$$

For cubic system

$$d_{\text{hkl}} = \frac{a}{\left(h^2 + k^2 + l^2\right)^{1/2}} = 350 \text{ pm}$$

$$= \frac{a}{\left(1^2 + 0^2 + 0^2\right)^{1/2}}$$

Hence a = 350 pm = 350×10^{-12} m

The density of crystal ρ is given by

$$\rho = \frac{nM}{NV}$$ where, n = no. of atoms in unit cell

M = molar mass of Li

V = volume of unit cell = a^3

N = Avogadro no. = 6.023×10^{23}

$$\therefore \quad n = \frac{\rho NV}{M} = \frac{\left(530 \text{ kg m}^{-1}\right)\left(6.023 \times 10^{23} \text{ mol}^{-1}\right)\left(350 \times 10^{-12} \text{ m}\right)^3}{6.94 \times 10^{-3} \text{ kg mol}^{-1}}$$

$$= 1.97 \simeq 2$$

The bcc unit cell having 8 atoms at 8 corners (contribution $8 \times \frac{1}{8}$ = 1) and one atom at the centre (contribution 1×1 = 1) has (1 + 1) two atoms per unit cell, *i.e.*, $x = 2$.

While *fcc* unit cell has 8 atoms at 8 corners (contribution $8 \times \frac{1}{8}$ = 1), one atom at each of the 6 faces (contribution $6 \times \frac{1}{2}$ = 3) shared by two unit cell, has (1 + 3) four atoms per unit cell, *i.e.*, $x = 4$.

Hence, Li metal crystallizes in bcc lattice.

3.10. STRUCTURE OF CRYSTALS

3.10.1 Structure of Rock Salt (NaCl) Crystal

The X-ray diffraction studies of sodium chloride crystal give the following results:

Table 3.1 Bragg's angle for various planes of NaCl crystal

Plane	*Order of Diffraction*					
	1st $(n = 1)$		2nd $(n = 2)$		3rd $(n = 3)$	
	θ_1	$\sin\theta_1$	θ_2	$\sin\theta_2$	θ_3	$\sin\theta_3$
100	5.9	0.103	11.9	0.208	18.2	0.312
110	8.4	0.146	17.0	0.292	–	–
111	5.2	0.091	10.5	0.182	–	–

Table 3.2 Relative intensities of maximum diffraction for NaCl

Order	*Intensities*		
	(100)	(110)	(111)
First	100	50.40	9.00
Second	19.90	6.10	33.10
Third	4.87	0.71	0.58
Fourth	0.79		2.82
Fifth	0.12	–	0.14

Considering only the first-order diffraction we get from the above data that

$$d_{100} : d_{110} : d_{111} = \frac{1}{\sin\theta_1} : \frac{1}{\sin\theta_2} : \frac{1}{\sin\theta_3}.$$

$$= \frac{1}{0.103} : \frac{1}{0.146} : \frac{1}{0.091} = 1 : 0.70 : 1.14$$

These values of ratios are nearly same as obtained for a unit cell of face centered lattice of a cubic cell (*i.e.*, $d_{100} : d_{110} : d_{111} = 1 : 0.704 : 1.154$).

Further when the intensities of diffraction maxima for different orders and planes are considered, we find that the intensities due to (100) and (110) planes fall off regularly, while for those of (111) plane, the first order is abnormally weak, the second order is quite strong, the third order is again weak and fourth order shows strong intensities.

These results indicate that alternate layer or planes of chloride ions ($Cl^{\ominus}$ ion) separated by sodium ions ($Na^{\oplus}$ ion) appear in (111) type of planes of face centered cubic lattice as shown below in Fig. 3.13.

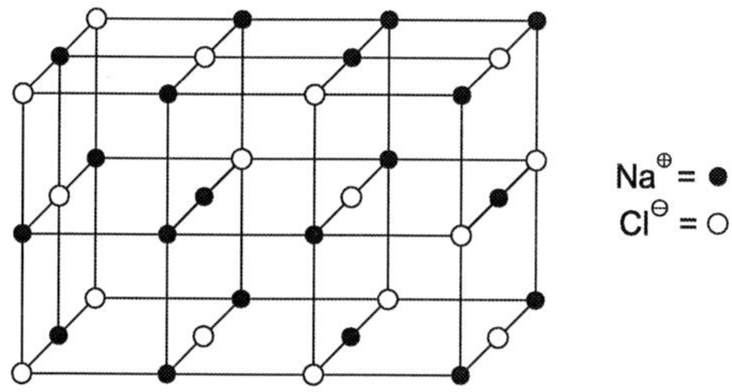

Fig. 3.13 Rock salt (NaCl) having fcc structure.

3.10.2 Structure of Potassium Chloride (KCl) Crystal

The first-order spectra from (100), (110) and (111) planes of KCl have been observed at 5.38°, 7.61° and 9.38°, respectively.

$$\text{Therefore} \quad d_{100} : d_{110} : d_{111} = \frac{1}{\sin 5.38} : \frac{1}{\sin 7.61} : \frac{1}{\sin 9.38}$$

$$= \frac{1}{0.0938} : \frac{1}{0.1326} : \frac{1}{0.1630} = 1 : 0.704 : 0.575.$$

These values of ratios are those found for simple cubic lattice. Hence KCl crystal has a simple cubic lattice but KCl is isomorphous with NaCl and must have *fcc* lattice. This discrepancy in X-ray diffraction can be explained by the fact that the scattering powers of $K^{\oplus}$ and $Cl^{\ominus}$ (both having identical extra nuclear structure or equal number of electrons) for X-rays is equal. The first-order diffraction from (111) planes are extinguished due to destructive interference and X-rays are showing results of simple cubic lattice structure rather than face centered cubic structure like NaCl. However, neutron diffraction studies of KCl crystal clearly show it to be *fcc*.

3.10.3 Structure of Caesium Chloride (CsCl) Crystal

The value of cube edge length of CsCl crystal as determined by Bragg's spectrometer is $3.510^{\oplus}$. The structure is body centered cubic with one molecule of CsCl per unit cell. Each $Cs^{\oplus}$ ion is surrounded by 8 $Cl^{\ominus}$ ions and each $Cl^{\ominus}$ ion is surrounded by 8 $Cs^{\oplus}$ ions so that co-ordination number of each ion is eight as shown in Fig. 3.14.

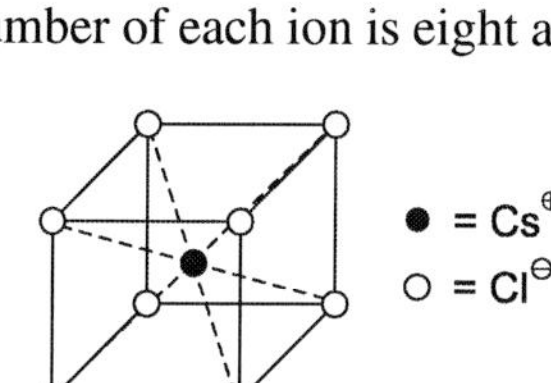

Fig. 3.14 CsCl crystal.

3.11 IMPERFECTION IN CRYSTALS

Real crystals are always imperfect in certain manners. When a crystal structure deviates from a periodic lattice, it is called an imperfection. Most common crystal imperfections are chemical impurities, vacant lattice sites and interstitial atoms. Trace amounts of chemical impurities are able to cause conductivity in certain semiconductors, while these imperfections are also able to impart colours to many crystals. Certain mechanical and plastic properties of solids are also controlled by these crystal defects.

3.11.1 Schottky Defect

When a crystal is having a lattice vacancy or hole *i.e.*, missing of an atom or ion in the crystal lattice as shown in figure below, we term it as Schottky defect.

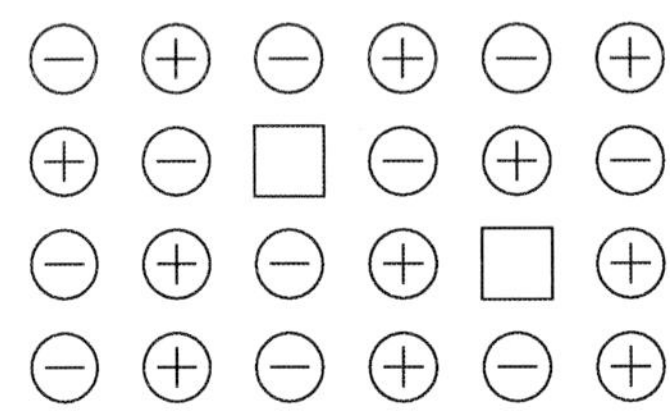

Fig. 3.15 Plane of the crystal of alkali halide showing a vacant positive and negative ion site (Schottky defect).

This defect is usually present in alkali halide crystals in which positive and negative ions are of similar size, *e.g.*, CsCl. This defect causes the lowering of the density of the crystal. It has been found that the mechanism of electrical conductivity in alkali halide is by motion of ions, *i.e.*, ionic in nature.

The number of Schottky defects (n) in a crystal containing N number of ions in crystal is

$$n = N \exp\left(\frac{-E}{2\,kT}\right) \text{ per cm}^3$$

where E is the energy necessary to create 1 mol of vacancy, k is the Boltzmann constant and T the absolute temperature.

3.11.2 Frenkel Defect

In this type of vacancy defect the missing ion or atom is transferred from lattice site to an interstitial position between the lattice planes:

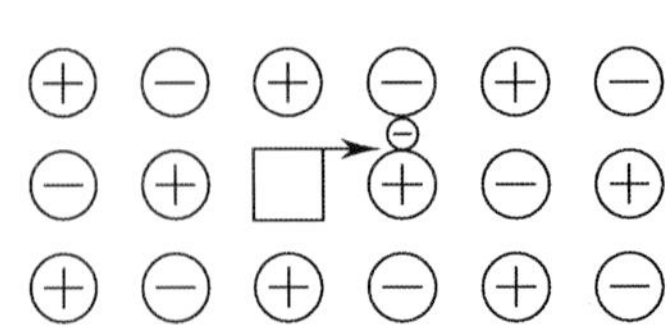

Fig. 3.16 Frenkel defect.

This defect is shown by silver halides, where some of the Ag^+ ions are generally missing from their regular positions and occupy position in between lattice planes, leaving the holes in its lattice. Frenkel defect is found usually in such compounds which have low co-ordination number and highly polarizing cation and easily polarizable anion, *e.g.*, AgBr.

The number of Frenkel defects in a crystal is given by

$$n = (N\,N_1)^{1/2} \exp\left(-\frac{E}{2kT}\right)$$

where E is the energy of formation of 1 mol of Frenkel defect, N the number of lattice sites, N_1 the possible interstitial sites at temperature T and k the Boltzmann constant.

3.11.3 Impurity Defects: Semiconductors

Addition of one boron atom to 10^5 silicon atoms, which is basically a non-conductor of electricity at ordinary temperature, increases the conductivity of pure silicon by 1000 times. Deliberate addition of impurities to such substances, like pure silicon, is called **doping,** wherein its electrical conductivity is enhanced many fold. The resulting material is termed as **semi-conductor.**

Let us consider the effect of impurities in silicon and germanium crystals. The elements crystallize in diamond structure as presented below:

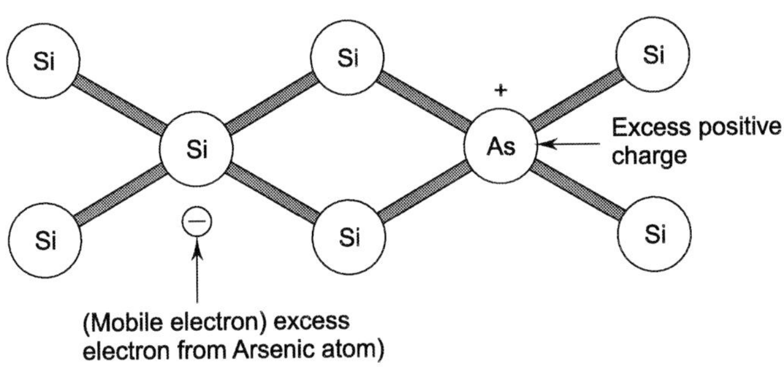

Fig. 3.17 *n*-type semiconductor.

Here each silicon atom is forming four covalent bonds and is attached with its four nearest neighbours. When an impurity atom, such as that of arsenic or antimony, with five valence electrons is substituted in the normal lattice of silicon atom, one of the valence electron left by the impurity atom after the establishment of four covalent bonds with nearest neighbours give a situation presented above, wherein the excess positive charge is provided on the arsenic atom when its mobile free electron is moved in the coulomb potential and the impuirity atom is thus called a **donor.** The crystal as a whole remains neutral. Similarly, when a trivalent impurity, *e.g.*, boron or gallium, called **acceptor** is introduced in silicon, a hole is found in the valence band, when one of the valence electron is taken up by the acceptor.

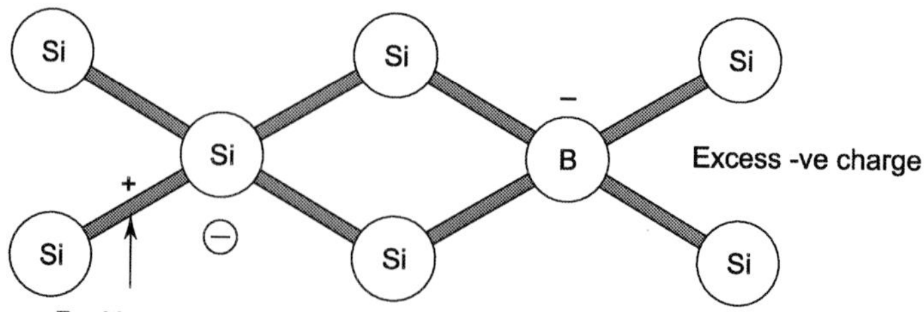

Fig. 3.18 *p*-type silicon semiconductor.

Semiconductor exhibiting conductivity due to flow of excess electrons (arsenic treated silicon crystals) are called *n*-type semiconductors, whereas those exhibiting electrical conductivity due to the formation of positive holes in the valence band (boron treated silicon crystals) are called *p*-type semiconductors. When a *n*-type semiconductor is joined through a *p*-type semiconducting material, a *p-n* junction device is obtained, and the same on exposure to sunlight can excite electrons from *n*-type semiconducting material to the holes of *p*-type. These are, therefore, able to furnish quite useful **solar-cell devices** for electric generators.

Objective Type Questions

1. Sodium metal belongs to which type of crystalline solid?
   (*a*) Ionic (*b*) Metallic
   (*c*) Covalent (*d*) It is amorphous.
2. Which of the following is not a characteristic of ionic solid ?
   (*a*) They are brittle.
   (*b*) They have high m.p.
   (*c*) They are soft.
   (*d*) They conduct electricity in fused state.
3. Cubic unit cell is characterized by
   (*a*) $a = b = c, \alpha = \beta = \gamma = 90°$ (*b*) $a = b = c, \alpha \neq \beta \neq c$
   (*c*) $a = b \neq c, \alpha = \beta = \gamma = 90°$ (*d*) $a = b \neq c, \alpha = \beta = 90°, \gamma \neq 90°$.
4. An example of triclinic unit cell is
   (*a*) Sulphur (*b*) $CaCO_3$
   (*c*) $Na_2B_4O_7 \cdot 10H_2O$ (*d*) $CuSO_4 \cdot 5H_2O$.
5. A cubic unit cell has atoms at the corner of the cube and in the centre of face of the cube. Number of atoms per unit cell is
   (*a*) 2 (*b*) 4
   (*c*) 8 (*d*) 14.

6. If the Weiss indices for plane are given by $2a : b : \alpha c$ then Miller indices for the same plane will be :

(a) $\frac{1}{2}:1:0$ (b) $\frac{1}{2}:1:\alpha$

(c) $2:1:0$ (d) $2:1:\alpha$.

7. The number of symmetry elements in a cubic crystal is

(a) 1 (b) 9

(c) 13 (d) 23.

8. Crystal structure may be determined experimentally by

(a) Bragg's method (b) Laue's method

(c) Powder method (d) All of these.

ANSWERS

1. (b) 2. (c) 3. (a) 4. (d) 5. (b)
6. (c) 7. (d) 8. (d).

Short Answer Type Questions

1. Classify each of the following solid as molecular, ionic, covalent or metallic

(a) CsCl (b) Na

(c) Diamond (d) Naphthalene.

[**Ans.** CsCl ionic, Na metallic, diamond covalent, naphthalene molecular]

2. A cubic unit cell has how many centre of symmetry, plane of symmetry and axes of symmetry ?

[**Ans.** 1 centre, 9 plane and 13 axes of symmetry]

3. Write Bragg's equation for obtaining lattice spacing between two successive planes of a crystal.

[**Ans.** Use $n\lambda = 2d \sin \theta$]

4. Determine the number of atoms in a unit cell of lithium which crystallizes in a body centered cubic arrangement.

$$\left[\textbf{Ans. } \text{1 in centre and 8 in corners or } (1 \times 1) + \left(8 \times \frac{1}{8}\right) = 2\right]$$

5. $LiBH_4$ has an orthorhombic structure with four molecules per unit cell. If the unit cell dimensions are $a = 6.8$, $b = 4.4$, and $c = 7.0$ Å, then calculate theoretical density of $LiBH_4$ (mol. weight : $LiBH_4 = 22\ g\ mol^{-1}$).

[**Ans.** 0.7 g/c.c.]

Subjective Questions

1. How many types of crystalline solids are known ? Give at least one example of each type.
2. Write brief note on :
   (a) Unit cell
   (b) Space lattice
   (c) Covalent solids.
3. Describe the Bravais lattices with proper diagrams.
4. Discuss the laws of crystallography.
5. What do you understand by Weiss and Miller indices ?
6. Give an account of the diffraction methods used for determination of crystal structure by monochromatic X-rays.
7. Barium metal has a body centered cubic lattice. Its density is 3.51 g/c.c. Calculate the edge length of the unit cell (given atomic Ba = 137 g mol^{-1} and $N_A = 6.023 \times 10^{23}$).

   [**Ans.** 5.06 Å]
8. Copper metal has a face centered cubic structure. The edge length of cubic cell is 3.61 Å. Calculate the radius of copper atom assuming it to be spherical.

   [**Hint :** spheres on the face of unit cell touch each other along the diagonal]

   [**Ans.** 1.28 Å]
9. What is understood by (100), (110) and (111) planes of cubic lattice ? How many such planes are there in a cube ?
10. What do you understand by spacing of successive planes ? How does the spacing of these differ in a cubic crystal ?
11. When X-rays of wavelength 1.54×10^{-10} m are used, the first reflection from KCl crystal occurs at 14° 12′. Determine the separation between lattice planes parallel to the face exposed to X-rays, and also the length of the unit cell; also give the density of KCl crystal.

    [**Ans.** $d_{100} = 3.139 \times 10^{-10}$ m, length of unit cell = 6.278×10^{-10} m, density of crystal = 2 g cm^{-3}]

    [**Hint:** (100) planes are exposed]

4

Colloidal State

4.1 INTRODUCTION

On the basis of diffusion studies, Graham classified substances as (*a*) cryztalloids, *e.g.*, sugar, salt, etc., and (*b*) Colloids, *e.g.*, glue, albumin, etc. It is now realized that there is no sharp distinction between crystalloids and colloids as the crystalloids also under suitable conditions can be converted into colloidal state.

In colloidal state there is a dispersion of particles of one substance (called **dispersed phase**) into another substance (called **dispersion medium** or continuous phase). Colloidal state thus differs from a true solution in being heterogenous as the dispersed particles are larger than those encountered in solution. It has been shown that the size of dispersed particles in collodial state is between 1 and 200 nm. If the particle size is less than 1 nm it becomes a true solution and if the particle size is above 200 nm it becomes a suspension.

Both the dispersed phase and dispersion medium may be solid, liquid or gaseous. Since mixture of two or more gases is always homogeneous, only eight types of colloidal dispersions are known as shown below in Table 4.1.

Table 4.1 Different types of colloidal dispersion

Dispersed Phase	*Dispersion Medium*	*Common (Class Name of Colloid)*	*Examples*
1. Gas	Liquid	Foam	Whipped Cream
2. Gas	Solid	Solid Foam	Dunlop Mattress
3. Liquid	Gas	Aerosol	Fog, Mist
4. Liquid	Liquid	Emulsion	Milk, Mayonnaise
5. Liquid	Solid	Gel	Fruit Jelly, Cheese
6. Solid	Gas	Aerosol	Smoke
7. Solid	Liquid	Sol	Albumin, Sulphur Sol
8. Solid	Solid	Solid sol	Ruby, Coloured Glass

4.2 CLASSIFICATION OF COLLOIDS

One classification of colloid is based on whether the dispersed phase and dispersion medium is gas, liquid or solid as shown in Table 4.1.

The most common type of colloidal system is one where dispersed phase is solid and dispersion medium is liquid. Such colloidal systems are known as sols. The sols are classified on the basis of dispersion medium. If the dispersion medium is water, they are called **hydrosols** or **aquasols.** If the dispersion medium is alcohol, they are called **alcosols** and if it is benzene then they are called **benzosols.**

Depending on the affinity of dispersed phase for dispersion medium, sols are also classified as :

(*i*) **Lyophilic Sols:** Lyophilic sols are colloidal systems where dispersed phase has strong affinity for dispersion medium. They are thermodynamically stable and easy to prepare. They are not easily coagulated by electrolytes and hence are also called **reversible colloids.** When the dispersion medium is water, they are also called hydrophilic sols. Examples are starch, gum, albumin, etc.

(*ii*) **Lyophobic Sols:** Lyophobic sols are colloidal systems where dispersed phase has little affinity for dispersion medium. They are generally unstable, easily coagulated by electrolytes and are difficult to prepare. For this reason they are referred to as **irreversible colloids.** If the dispersion medium happens to be water then they are called hydrophobic sols. Examples are all metal sols, sulphur sol, etc.

4.3 PREPARATION OF COLLOIDAL SOLS

Lyophilic sols are easily obtained by bringing the dispersed phase in contact with dispersion medium. However, lyophobic sols cannot be prepared like this and special methods are employed to obtain such sols. These methods either start from a bigger particle (suspension or solid state) and convert it to colloidal state **(dispersion methods)** or start from smaller particle (as in solution) and condense them to colloidal state **(condensation methods)**.

4.3.1 Dispersion Methods

(*i*) **Mechanical or Electrical Dispersion:** By grinding a solid in dispersion medium mechanically to get a colloidal sol. The device used is called **colloidal mill.** Preparation of ayurvedic medicines by grinding and preparation of milk shake in blender are common examples.

(*ii*) **Ultrasound Dispersion:** By exposing the dispersed phase to ultrasound. Sol of mercury in water is prepared by this method.

(*iii*) **Bredig's Arc Method:** This method is used to prepare sols of metals like Ag, Au, etc. It consists of striking an arc between electrodes of metal, whose sol is to be prepared and immersed in ice-cooled dispersion medium.

(*iv*) **Peptization:** Sometimes a freshly obtained precipitate in presence of a common ion passes into colloidal state. The phenomenon is known as peptization. For example, freshly precipitated ferric phosphate gives a colloidal sol in presence of ferric ions.

4.3.2 Condensation Methods

(*i*) **Excessive Cooling:** Those substances whose solubility decreases with decrease in temperature can be converted to colloidal state by excessive cooling of their nearly saturated solution. Cooling of sulphur solution in ethanol forms its colloidal sol.

(*ii*) **Exchange of Solvent:** If a substance is soluble in one solvent but insoluble in another then it can be converted to colloidal state by mixing the solution of substance with other solvent. For example, sulphur is slightly soluble in alcohol but insoluble in water. Sulphur sol can be prepared by pouring drops of solution of sulphur in alcohol into water.

(*iii*) **Chemical Methods :**

(*a*) **Oxidation:** Colloidal sol of sulphur may be prepared by passing H_2S gas in a dilute solution of HNO_3 when H_2S is oxidized to sulphur.

$$H_2S + [O] \longrightarrow H_2O + S$$

(*b*) **Reduction:** Purple of Cassius a co-colloid of Au and stannic acid is obtained by reducing a solution of $AuCl_3$ with $SnCl_2$.

$$2AuCl_3 + 3SnCl_2 \longrightarrow 2Au + 3SnCl_4;$$

$$SnCl_4 + 4H_2O \longrightarrow Sn\,(OH)_4 + 4HCl.$$

(*c*) **Hydrolysis:** Ferric hydroxide sol can be obtained by pouring a solution of ferric chloride in water.

$$FeCl_3 + 3H_2O \longrightarrow Fe\,(OH)_3 + 3HCl$$

(*d*) **Double Decomposition:** Sol of arsenious sulphide can be obtained by passing H_2S intro a saturated solution of arsenious oxide.

$$As_2O_3 + 3H_2S \longrightarrow As_2S_3 + 3H_2O$$

4.4 PURIFICATION OF COLLOIDAL SOLS

Lyophobic colloidal sols prepared as above are generally contaminated by the presence of electrolytes which tend to coagulate these sols. Therefore, it is necessary to remove them. Common methods used for purification of such sols are : (*i*) dialysis; (*ii*) electrodialysis ; and (*iii*) ultrafiltration.

(*i*) **Dialysis:** The sol is taken in a bag of suitable membrane (*e.g.*, cellophane, parchment, etc.) which is then suspended in a tank of water having a running water supply. The process is based on the fact that the membrane allows smaller electrolyte particles to pass through it, while colloidal particles being bigger in size cannot.

(*ii*) **Electrodialysis:** In order to hasten the removal of electrolytes an electric field is applied across the membrane in dialysis. The ions move faster in the electric field and are removed easily. The process is commonly used for purification of blood in patients with kidney failure. Such a dialyser is called artificial kidney machine.

(*iii*) **Ultrafiltration:** The process is similar to ordinary filtration except that the filtration membrane used here is impregnated with suitable material so that it permits the passage of only electrolytes and dispersion medium but not the dispersed phase. The membrane used is called ultrafiltre and is more useful in the purification of lyophilic sols.

4.5 PROPERTIES OF COLLOIDAL SOLS

Since the particle size of the dispersed phase is much bigger than that of solute in solution, the colloidal sols are heterogenous in nature. The values of colligative properties like elevation of boiling point, depression of freezing point and osmotic pressure are much lower than those of true solution. Osmotic pressure measurements are often used to measure molecular mass of dispersed phase which gives information regarding number of particles per unit weight of dispersion medium.

The particle size of colloidal sol is much smaller than solid suspension or precipitate which results in increased surface area or interfacial area. As a result, the adsorption capacity of such a sol is much higher. Many colloidal sols are, therefore, used as effective catalysts in many industrial processes.

For lyophobic sols the values of surface tension, viscosity, etc., are nearly the same as those of dispersion medium but the viscosity measurements of lyophilic sols can be used to get information about their molecular mass.

Other properties of colloidal sols may be grouped under following three headings :

(*i*) **Optical Properties:** Colloidal sols are transluscent in appearance. They may be colourless or coloured. It is interesting to note that the colour of a sol is dependent on the size of the dispersed phase rather than on its nature. As the size becomes bigger the colour changes from red to blue to violet.

(*a*) **Faraday–Tyndall Effect:** When a beam of light is passed through a colloidal sol, the path of the light beam gets illuminated because of the scattering of light by dispersed phase. The effect was first observed by Tyndall and is known as Faraday-Tyndall effect or simply as Tyndall effect. It is observed that the zone of scattered light is much larger than the particles of dispersed phase.

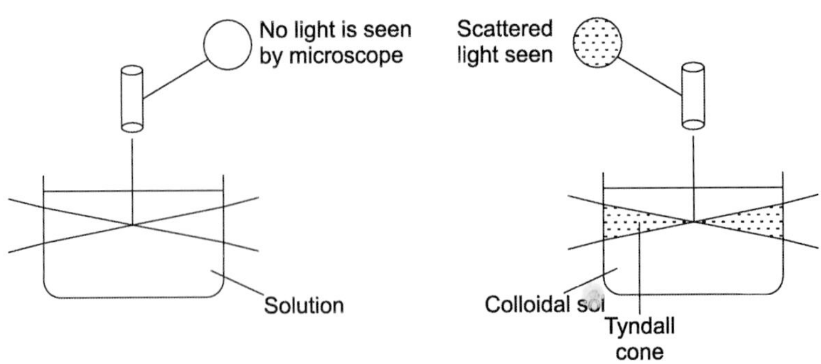

Fig. 4.1 Difference between solution and colloidal sol.

The effect was used by Zsigmondy to devise **ultra-microscope** which can be used to determine size, shape and molecular mass of dispersed phase. If 'a' gram of substance is present per mL, and 'n' number of particles are present in V mL (detected by ultra-microscope), then assuming that the colloidal particle is spherical, its radius 'r' can be calculated by

$$r = \sqrt[3]{\frac{3aV}{4\pi n \rho}}, \text{ where } l = \text{density of dry dispersed phase.}$$

(*ii*) **Kinetic Properties:** Colloidal sols when observed under a microscope show a continuous random motion. The particles move in zig-zag short straight paths of different lengths in different directions. The phenomenon was first observed by Robert Brown while studying pollen grains and is called **Brownian motion.**

Werner pointed out that Brownian motion was caused by continuous bombardment of dispersed phase by molecules of dispersion medium.

The mean square displacement $\langle x^2 \rangle$ of dispersed phase in time 't' can be calculated using Einstein equation

$$\langle x^2 \rangle = t\frac{kT}{3\eta\pi r}$$

where, k is Boltzmann constant T is temperature in Kelvin; η is viscosity of the medium ; and r is radius of dispersed phase.

(*iii*) **Electrical Properties:** All sol particles in a dispersion medium carry electrical charges. Some of these are positively charged while others are negatively charged. Thus ferric hydroxide sol is positively charged and arsenious oxide sol is negatively charged. Some sols may be positively or negatively charged depending on the method of preparation.

(*a*) **Electrophoresis:** Migration of dispersed phase towards the electrodes under the action of electric field is called electrophoresis. If the particles are positively charged and move towards cathode, the phenomenon is called **cataphoresis** but if the particles are negatively charged and move towards anode, it is called **anaphoresis**.

The velocity of dispersed phase under a potential of 1 volt per cm is called its **electrophoretic mobility**. Electrophoretic mobility may be calculated using the following equation :

$$v = \frac{ZD}{4\pi\eta}$$

where, Z =Zeta potential; D = dielectric constant of solvent η = viscosity of medium; and v = electrophoretic mobility.

(*b*) **Electro-osmosis:** When an electric field is applied to a colloidal sol and dispersed phase is not allowed to move by a physical barrier, then it is observed that dispersion medium moves in a direction opposite to that of dispersed phase. This phenomenon of migration of dispersion medium under the influence of electric field is called electro-osmosis.

4.6 ORIGIN OF CHARGE ON COLLOIDAL SOLS

The origin of charge on colloidal sols may be due to one or more of the following reasons :

(*i*) **Self-ionization:** Many colloidal sols contain groups which readily ionize in dispersion medium giving charge to the dispersed phase. Most common example is detergents and common soap (RCOONa) which readily ionize to form groups of $RCOO^{\ominus}$ known as **micelle**. Protein sols also contain ionizable acidic as well as basic groups which make them either positively or negatively charged depending on the pH of the medium.

(*ii*) **Selective Adsorption of Ions:** Due to greater surface area the adsorption capacity of colloidal sols is high. The charge on many colloidal sols is due to selective adsorption of positive or negative ions present in the medium. The dispersed phase acquires the charge of adsorbed ion. Thus in ferric hydroxide sol if there is an excess of $Fe^{3\oplus}$ ions, these are adsorbed and dispersed phase gets a positive charge but in presence of excess $OH^{\ominus}$ ions, these are adsorbed and dispersed phase gets a negative charge.

(*iii*) **Helmholtz Electrical Double Layer Theory:** According to this theory preferential adsorption of an ion (positive or negative) by colloidal sol particles imparts a positive or negative charge to it. In the vicinity of dispersed phase there is an excess of **counter ions** (ions of opposite charge). The force of attraction between them is sufficiently strong so that counter ions are practically immobilized forming a fixed layer known as **Stern layer** (first layer). The other part of double layer consists of a **mobile layer** of opposite ions which extends into the bulk of the medium. This layer consists of ions of opposite charge as well as same charge. However, the net charge on mobile layer is equal and opposite to that of fixed layer.

Let us consider the fomation of silver iodide sol from double decomposition reaction.

$$AgNO_3\ (aq) + NaI\ (aq) \rightarrow AgI + NaNO_3\ (aq)$$

When excess of sodium iodide is present, it would result in the preferential absorption of I–ions giving a negatively charged sol of silver iodide:

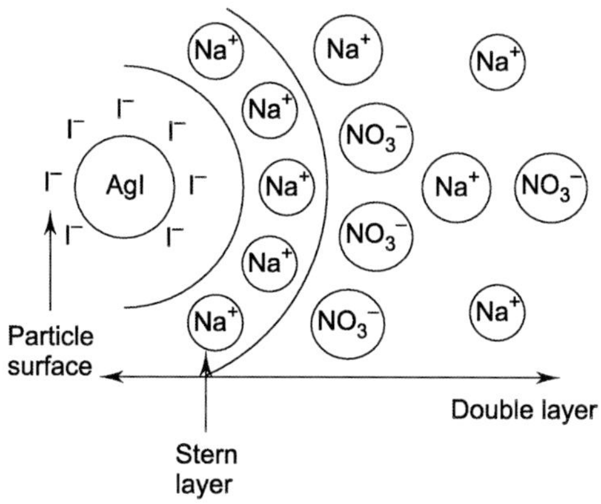

Fig. 4.2 Formation of AgI sol.

The positive Na^+ ions are present near the stern layer and are called counter ions, while iodide ions adsorbed on the surface of sol particle are the potential-determining ions.

Due to the presence of opposite charges on fixed and mobile layer there is a difference of potential between the two, known as **zeta potential,** which is responsible for the stability of colloidal sol.

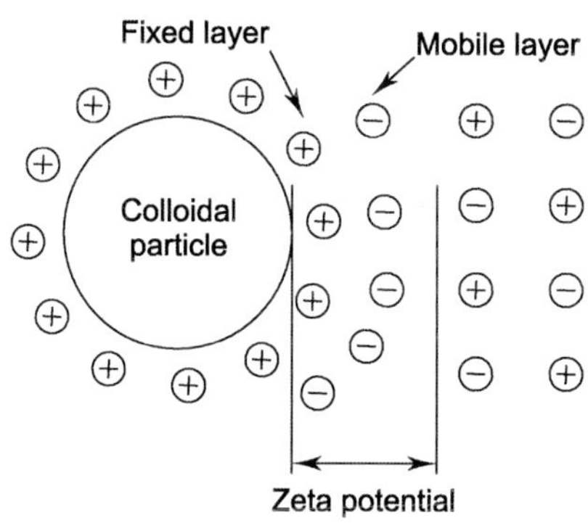

Fig. 4.3 Helmholtz double layer on colloidal particle.

Zeta potential may be calculated using equation given under electrophoresis, *i.e.*,

$$v = \frac{ZD}{4\pi\eta}$$

Table 4.2 Zeta potential of some aquasols

Dispersed Phase	*Dispersion Medium*	*Zeta Potential*
Arsenious Sulphide	Water	– 0.032 volts
Gold	Water	– 0.058 volts
Silver	Water	– 0.033 volts
Ferric Oxide	Water	+ 0.044 volts

4.7 COAGULATION OF COLLOIDS

Colloidal sols may be precipitated by various means like physical methods (like stirring, cooling and boiling) or addition of electrolytes and dehydrating agents or change of dispersion medium. The phenomenon of precipitation of colloidal sol is commonly known as **coagulation** or **flocculation**.

Lyophilic sols are generally stable and are not coagulated easily. This is because of strong interaction between dispersed phase and dispersion medium leading to solvation of dispersed phase.

When an electrolyte is added to lyophilic sol, the ions of the electrolyte compete with dispersed phase for getting solvated. Hence, large amount of electrolyte is required to coagulate the sol. This is often called salting out of sol.

Alternatively addition of a dehydrating agent or addition of a competing solvent which displaces the solvated dispersion medium, leads to coagulation.

Lyophobic sols on the other hand are much more sensitive to addition of electrolytes. Normally the particles of disperse phase do not coalesce due to the presence of electrical double layer of same charge which acts as a potential barrier in not allowing them to collide with each other.

When an electrolyte is added to the sol the concentration of co-ions (ions of same charge) and counter ions (ions of opposite charge) increases which leads to decrease in the thickness of double layer and consequently decrease in the zeta potential also. Therefore, the particles of dispersed phase can come nearer, collide with each other and collapse to form a bigger particle leading to coagulation.

Hardy and Schulze studied the coagulation of sols by electrolyte and formulated a rule known as **Hardy-Schulze rule** which states (*i*) the ion which is effective in causing coagulation of sol (known as counter ion) is the one with charge opposite to that of dispersed phase **(counter-ion),** and (*ii*) greater the charge of counter ion, greater is the **coagulation power** or **flocculation power** of the electrolyte.

Thus for a negatively charged sol, the flocculation power of counter ions will be in the order $Na^{\oplus} < Ba^{2\oplus} < Al^{3\oplus}$ while for a positively charged sol the flocculation power of counter ions will be in the order $Cl^{\ominus} < SO_4^{2\ominus} < PO_4^{3\ominus}$.

Hofmeister has found that even amongst the monovalent negative ions the flocculation power is in the order:

$$CH_3COO^{\ominus} < Cl^{\ominus} < Br^{\ominus} < NO_2^{\ominus} < I^{\ominus}$$

The above series is known as **lyotropic series.** Sometimes another term **flocculation value** is used to define coagulating properties of an electrolyte. Flocculation value is defined as the ***minimum number of milli-moles of electrolyte required for coagulation of one litre sol.*** Obviously the order of flocculation value is just the reverse of flocculation power. The electrolyte which has high flocculation power must have low flocculation value.

4.8 PROTECTION OF COLLOIDAL SOLS

It should be quite obvious by now that lyophobic colloids are difficult to prepare but are easily coagulated. It may sometimes be necessary to keep these colloidal sols for longer period of time. Hence, there is a need to protect these colloids from coagulation.

The process of protection is essentially a step to neutralize the forces and conditions which are responsible for coagulation. If a very small amount of lyophilic sol is added to a lyophobic sol–it is found that the latter becomes less sensitive to coagulation by electrolytes. This low **sensitization** towards coagulation is perhaps due to adsorption of oppositely charged lyophilic particles by the lyophobic sol.

If the lyophilic sol has the same charge as the lyophobic sol and the former is added in relatively sufficient quantity, then the lyophobic sol is generally prevented from coagulation on addition of an electrolyte.

The added lyophilic sol is called **protective colloid** and the process of stabilization of lyophobic sol is called **protection**.

The substances commonly used as protective colloids are all hydrophilic sols like gelatin, glue, starch, albumin, etc.

Zsigmonly introduced the term **gold number** to describe the protecting power of many protecting colloids. Gold number is defined as,

"The number of milligrams of the protective colloid that must be added to 10 mL of the standard gold sol so that addition of 1 mL 10% NaCl solution just prevents its coagulation".

Almost near coagulation of gold sol is indicated by a change of colour from red to blue.

The value of gold numbers of some protective colloids is given in Table 4.3.

Table 4.3

Protective Colloid	Gold Numbers of Some Protective Colloids
Gelatin	0.005 – 0.01
Haemoglobin	0.03 – 0.07
Gum arabic	0.15 – 0.5
Albumin	0.1 – 0.2
Starch	~ 25

It may be noted that smaller the gold number greater is the protective power of lyophilic colloid. Protective action of colloids is very useful in many industrial processes like making of ice-cream, use of Ag sols as medicine and in silver mirroring, etc.

Another alternative method of measuring protective power of colloids is **Congo-rubin** number which uses congo red sol instead of gold sol.

4.9 EMULSIONS

The colloidal systems in which both dispersed phase and dispersion medium are liquids, are called emulsions–common examples being milk, mayonnaise and cod liver oil. In most emulsions one liquid phase is water and the other liquid is immiscible with water commonly an oil or fat. As a consequence there are two types of emulsions : (*i*) oil in water, *e.g.*, milk and (*ii*) water in oil, *e.g.*, cod liver oil.

The particles of dispersed phase in emulsions are larger than those of sols and can be viewed under microscope. In order to prepare emulsion of first type, *i.e.*, oil in water, the oil or fat is shaken with water or distilled in steam when a faint milky emulsion is obtained.

Water in oil emulsions usually exist only in the presence of a stabilizer because the interfacial tension between water and oil is generally large. In order for such emulsion to be stable there should be a large area of oil-water interface otherwise the globules of dispersed phase will coalesce and coagulate the emulsion. Substances which are used to stabilize emulsions are known as emulsifiers.

4.9.1 Emulsifiers

Most common emulsifiers are long chain compounds with polar groups, *e.g.*, soaps, detergents (alkyl sulphonates, etc.) which appreciably reduce oil-water interfacial tension. The cleaning action of soap is due to its ability to emulsify oil.

Other emulsifiers include lyophilic sols such as proteins, gum and some insoluble substances, *e.g.*, $PbSO_4$, sulphates of Fe, Cu and lamp black, etc.

It has also been observed that an oil in water emulsion on addition of common soap as emulsifier may be converted into water in oil emulsion on extra addition of salts of di- and trivalent cations, *e.g.*, Al^{3+}, Fe^{3+}, Ni^{2+}, Pb^{2+}, etc. This is probably due to the formation of a soap of these metal ions.

An emulsifier like soap has a oil soluble hydrocarbon group and water soluble polar group. At the oil-water interface the polar group is oriented towards water and hydrocarbon group towards the oil. If the cross section of polar end is larger than the hydrocarbon end then water will be a continuous phase and will favour the formation of oil in water emulsion.

In case of soaps of multivalent metal ions, it is expected that the cross section of hydrocarbon end will be larger, making it the continuous phase and favours the formation of water in oil emulsion. This appears to be the reason for phase reversal on addition of polyvalent metal ions.

4.10 GELS

A colloidal system in which liquid is the dispersed phase and solid the dispersion medium are called gels, *e.g.*, gelatin, silica gel, etc. Gels may be considered to have properties which are intermediate between those of solids and sols. Thus, they are colloidal systems in which each component is continuous throughout the system. Gels like gelatin, agar-agar have water as the dispersed phase while soft paraffin has an organic liquid as the dispersed phase.

Depending on their physical properties gels may be classified as :

(*i*) **Elastic Gels:** As the name suggests they are elastic in nature and their shape is altered on application of force but the original shape is restored when the force is withdrawn. Partial dehydration of an elastic gel results in the formation of an elastic solid which readily changes to original gel when brought in contact with liquid. In elastic gel, the phases are held by electrostatic forces. Common examples of elastic gel are gelatin, fruit jam and jellies, etc.

(*ii*) **Rigid or Non-elastic Gels:** Unlike elastic gels, nonelastic gels lose their elasticity on drying and gel cannot be restored when brought in contact with liquid. The rigid structure of these gels is due to a network of chemical bonds. Common examples are silica gel, alumina gel etc., used as dehydrating agents.

The main difference between elastic and rigid gels is perhaps the rigidity of the walls of capillary network in solid dispersion medium. In elastic gels, these capillaries are supple while in rigid gels they are rigid and honeycombed.

(*iii*) **Thixotropic Gels:** Gels which acquire the property of flow on application of pressure but again become semi-solid when at rest (like toothpaste) are called thixotropic gels. This property is attributed to the presence of weak attractive forces between phases. Common example is a gel of kieselquhr and water, bentonite and water, ferric oxide and water, etc.

4.10.1 Properties of Gels

(*i*) **Imbibition or Swelling :** When placed in contact with a suitable liquid that solvates it, the gel is able to imbibe significant volume of the liquid resulting in swelling. The phenomenon is called **imbibition.**

The extent of swelling is dependent on the number and strength of linkages between individual molecules of gelling agent and is often several times the original volume. Imbibition also depends on the pH of the solvent and for a protein gel the swelling is minimum at its isoelectric point. When swelling is significant, the gel often disperses and forms a sol spontaneously or on stirring or warming.

(*ii*) **Syneresis and Bleeding :** Loss of fluid by a gel on standing is called **bleeding.** Sometimes this loss of fluid is accompanied by shrinking of the gel when it is termed as **syneresis.** It may be noted that neither bleeding nor syneresis is reverse of imbibition as there is no change in volume in syneresis. Syneresis is believed to be due to loss of fluid held by capillary forces by the frame work of gel structure. The degree of syneresis decreases as the concentration of gelling agent increases. Syneresis is shown by silicic acid gels, agar-agar gels and blood clot, etc.

(*iii*) **Ageing :** Most gels on standing tend to acquire a denser network of gelling agent leading to aggregation and increase in hardness. The phenomenon is called **ageing.**

(*iv*) **Formation of Liesegang Rings :** If a drop of silver nitrate solution is placed on a thin film of gelatin containing potassium dichromate, then a number of concentric rings of coloured silver chromate known as Liesegang rings are formed. Such rings are also observed if the gel is agar-agar or silicic acid.

4.11 MICELLE FORMATION

When molecules having both a hydrophilic and a hydrophobic group are dispersed in water to form a colloidal sol, they tend to associate to form colloidal particles called **micelle**.

Thus, micelle is a colloidal dispersed phase formed by the association of molecules in such a manner that hydrophobic ends point towards one another while the hydrophilic ends point outward facing the dispersion medium water.

Common washing soap is sodium stearate $C_{17}H_{35}COONa$. The long chain alkyl group is hydrophobic while –COONa group ionizes to give $COO^{\ominus}$ which is a hydrophilic end. In water solution stearate ions associate to form micelle such that hydrocarbon ends point towards centre while carboxyl groups point outward.

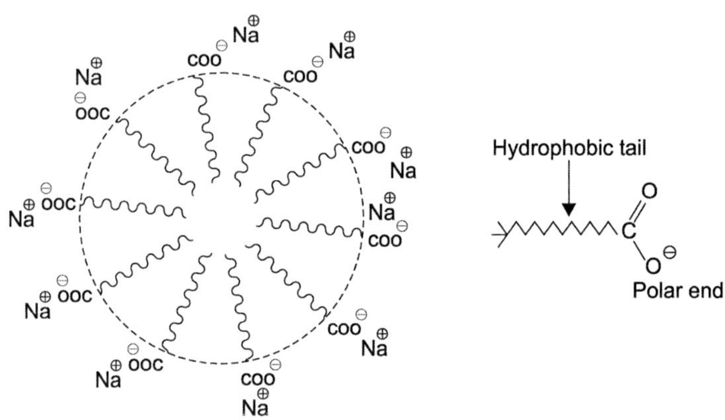

Fig. 4.4 Showing micelle formation by soap.

When dirty laundry is soaked in soap solution, the grease or oil gets absorbed towards the centre in the hydrophobic end. On beating or stirring the soap along with grease or oil settles down thus cleaning the laundry item.

Synthetic detergents are made based on the model of soap and also form micelle in solution. Common example of anionic detergent is sodium dodecylbenzenesulphonate and that of cationic detergent is dimethyl distearyl ammonium bromide.

4.12 APPLICATIONS OF COLLOIDS

Colloids find a multitude of usage in day to day life, as well as in industry, medicine, food stuff, biology, environment, etc. Some of the important applications are given below :

(*i*) Colloids find great importance in biology and living systems. Blood is a colloidal sol while protoplasm is a gel. Bile juice is used in the body to form oil–water emulsion. Most of the hormones are released in the body as colloidal sol. Many of the proteins like albumin, etc. are colloidal in nature.

(*ii*) Colloids are often used in pharmaceuticals and medicines as their smaller size and greater adsorption capacity makes them easily assimilable and more effective. Common examples are milk of magnesia used to control stomach acidity, iron and calcium tonics, formulations of antibiotics and other drugs, many injectible solutions, colloidal Ag, Au ointments, etc.

(*iii*) **Colloids in industry and environment**

(*a*) **Cottrell's** smoke precipitator. Smoke coming out of chimneys used in industrial processes is a colloid which contains many poisonous and irritant substances. These are removed by Cottrell's precipitator in which smoke is passed over high voltage electric plates. As a result the charged colloidal particles of coal and other pollutants are precipitated which reduces the air pollution.

(*b*) Leather industry uses colloidal dispersions of tannin, chromium, aluminium, etc. in tanning of leather.

(*c*) Textile industry uses colloidal dispersions of mordants and dyes to colour textiles.

(*d*) Paint industry produces paints, varnishes, etc., which are essentially colloidal in nature.

(*e*) Soap, detergents and other cleaning agents are colloidal substances used to remove grease, oil and dirt. Cleaning action of soap and detergents is due to their emulsifying properties.

(*f*) Purification of water is done by precipitating the colloidal impurities by alum.

(*g*) Sewage water contains colloidal dispersion of organic matter, mud, sewage, soap, etc. In sewage treatment this colloidal matter is removed by coagulation, adsorption and filtration.

(*h*) Fire extinguishers produce a froth – a colloidal foam by reaction of sodium bicarbonate and alum.

(*i*) In nature mud, cloud, fog, mist are various kinds of colloids. Rubber is a colloid which is deposited on coat, gloves to get rubberized aprons, gloves, etc.

(*j*) Delta formation at the termination point where rivers meet the sea is due to precipitation of colloidal clay and sand by electrolytes present in sea water.

(*iv*) Majority of items of food stuff are colloidal in nature. Some examples are milk, cheese, butter, curd, fruit jellies and jams, vegetable gravy, milk and fruit shakes, ready to drink tea and coffee, fruit juices, some cold drinks using brominated vegetable oil for emulsification, etc.

(*v*) Amongst other substances of daily use colloids are often encountered in the form of tooth paste, shampoo, creams, shaving foam, scent spray, room freshners, insect spray, disinfectant spray, etc.

Objective Type Questions

1. In the colloidal state the size of dispersed phase particle is in the range

(*a*) 1–200 Å (*b*) 1–200 mμ
(*c*) 1–200 nm (*d*) 100–200 Å.

2. Colloidal sols can be purified by

(*a*) Dialysis (*b*) Coagulation
(*c*) Flocculation (*d*) Peptization.

3. Which of the following system is not a colloidal state ?

(*a*) Gas in liquid (*b*) Gas in gas
(*c*) Gas in solid (*d*) Solid in gas.

4. The dispersion medium in benzosol is generally :

(*a*) Water (*b*) Alcohol
(*c*) Benzene (*d*) Benzpyrene.

5. When a strong beam of light is passed through a colloidal sol, it is

(*a*) Reflected (*b*) Dispersed
(*c*) Absorbed (*d*) Scattered.

6. Migration of dispersion medium of a colloidal sol in applied electric field is called

(*a*) Electrophoresis (*b*) Brownian motion
(*c*) Tyndall effect (*d*) Electro-osmosis.

7. The role of emulsifier is to

(*a*) Stabilize emulsions (*b*) Coagulate emulsions
(*c*) Convert it to true solution (*d*) Convert it to a gel.

8. The flocculation power of which of ions is maximum for coagulation of a negatively charged sol

(*a*) $Na^{\oplus}$ (*b*) $Ba^{2\oplus}$
(*c*) $Fe^{3\oplus}$ (*d*) All have same power.

9. Which of the following is not used as a protective colloid ?

(*a*) Gelatin (*b*) Silica gel
(*c*) Albumin (*d*) Starch.

10. In gels the dispersed phase and dispersion medium are respectively

(*a*) Liquid and solid (*b*) Solid and liquid
(*c*) Liquid and liquid (*d*) Solid and gas.

ANSWERS

1. (*c*)	**2.** (*a*)	**3.** (*b*)	**4.** (*c*)	**5.** (*d*)
6. (*d*)	**7.** (*a*)	**8.** (*c*)	**9.** (*b*)	**10.** (*a*)

Short Answer Type Questions

1. Explain the following terms in one line each :
   (*a*) Imbibition (*b*) Syneresis

2. Why gas-gas colloidal system does not exist ?

3. Give two examples each of
   (*a*) Foam (*b*) Emulsion
   (*c*) Sol (*d*) Gel.

4. Describe in brief Bredig's are method for obtaining metal sols.

5. Name two methods used for purification of colloidal sols.

6. Mention the name of one property each of colloidal sol belonging to the below category
   (*a*) Optical property (*b*) Kinetic property
   (*c*) Electrical property.

7. (*a*) What is Hardy-Schulze rule ?
   (*b*) Arrange NaCl, $CaCl_2$, $FeCl_3$ in order of increasing flocculation value.

8. Define gold number.

9. Explain in brief the formation of soap micelle.

10. Arrange the following ions in order of increasing flocculation power $Ca^{2\oplus}$, $K^{\oplus}$, $Al^{3\oplus}$.

Subjective Questions

1. What are colloids ? How many types of colloidal dispersions are known ?

2. Explain the difference between reversible and irreversible colloids.

3. Give a brief account of dispersion and condensation methods for the preparation of colloids.

4. Which method would you prefer to obtain a gold sol starting from metallic gold and water ?

5. Compare the properties of colloidal sols with true solutions taking into consideration the following aspects :
   (*a*) Osmotic pressure (*b*) Action of strong beam of light
   (*c*) Depression in freezing point.

6. Write brief notes on:

(*a*) Ultra–microscope (*b*) Brownian motion

(*c*) Electrophoresis.

7. Write a brief note on origin of charge on colloidal particles with special emphasis on Helmholtz double layer theory.

8. Explain the mechanism of coagulation of a lyophobic sol when an electrolyte is added to it. What is the role of protective colloid in stabilizing a lyophobic colloid?

9. How many types of emulsions are known? What is the role of emulsifiers in stabilizing emulsions?

10. What are gels? How many classes of gels are known? Discuss in brief the properties of gels.

11. A silver sol in water contains 0.2 g silver per cm^3. The number of particles of dispersed phase is 4.5×10^{12} per cm^3 of colloidal sol.

Assuming silver sol particles to be spherical, calculate the radius of the particle. Given density of metallic Ag = 10.5 g cm^{-3}.

[**Ans.** 120 nm]

❑❑❑

5

Chemical Kinetics and Catalysis

5.1 INTRODUCTION

Whether a reaction will occur under a given set of conditions can be predicted by thermodynamics. Even when thermodynamics predicts that a certain reaction is feasible, but if it takes say hundred years for the reaction to occur, there is no practical significance of such a reaction. Hence, the rate at which a reaction occurs acquires immense significance.

All reactions do not occur at same speed, some like corrosion are very slow while others like precipitation of AgCl by addition of HCl to $AgNO_3$ are almost instantaneous. Chemical kinetics is the study of reaction rates, their measurements, interpretations, effect of factors that influence the rate of reaction, *e.g.*, operating conditions, temperature, *etc.*, and if something can be done to change the rate of a particular reaction. Thus, it provides the fundamental knowledge for reactor design, scale-up optimization, operation and control which are key tasks of building and operating a chemical plant or petroleum refinery.

Kinetics thus helps in the study of reaction mechanisms and development of new industrial processes.

5.2 RATE OF CHEMICAL REACTION

The rate of a chemical reaction may be defined as the change in concentration of any reactant or product of reaction in a particular time interval. For a reaction of the type

$$A + B \longrightarrow C$$

if change in concentrations of A, B and C are $d[A]$, $d[B]$ and $d[C]$ in time dt, then rate of reaction may be written as

$$\text{Rate of reaction} = -\frac{d[A]}{dt} = -\frac{d[B]}{dt} = \frac{d[C]}{dt}$$

Obviously the concentration of A and B will decrease while that of C will increase. Thus, negative sign shown in case of reactants A and B indicates decrease in their concentration with time and a positive sign in case of product indicates increase in concentration with time.

The rates of reactions expressed in terms of reactants or products are related to each other by the stoichiometry of the reaction. Thus for the hypothetical reaction

$$2A + B \longrightarrow 3C$$

Since two moles of $[A]$ combines with one mole of $[B]$, rate of decrease in concentration of $[A]$ must be twice that of $[B]$ and consequently, the rate of increase in concentration of $[C]$ must be three times the decrease in concentration of $[B]$. Thus, the rates may be expressed in terms of reactants or products as follows:

$$-\frac{1}{2}\frac{d[A]}{dt} = -\frac{d[B]}{dt} = \frac{1}{3}\frac{d[C]}{dt}$$

For a general reaction,

$$xA + yB \rightleftharpoons nC + mD$$

the rates are expressed conventionally as :

$$-\frac{1}{x}\frac{d[A]}{dt} = -\frac{1}{y}\frac{d[B]}{dt} = \frac{1}{n}\frac{d[C]}{dt} = \frac{1}{m}\frac{d[D]}{dt}$$

Thus, the rate of reaction for any species A is defined as

$$\text{rate} = \frac{1}{v_A}\frac{d[A]}{dt}$$

where, v_A is the stoichiometric coefficient of species A in the balanced chemical equation. In addition the stoichiometric coefficients are given negative signs for the reactants and positive signs for the products. This definition ensures that the rate of a reaction is always positive and same for a given reaction no matter which species is considered.

According to law of mass action, the rate at which a substance reacts is proportional to its active mass. The active mass of reactants continuously goes on decreasing as the reaction progresses thus continuously changing the rates of reactions. Hence, it must be borne in mind that either we talk of **initial rate,** *i.e.*, the rate at the start of the reaction or **average rate,** *i.e.*, average of rate over a time period. Because the rate varies with time (and concentration) it is usual to define the rate over a very small time, Δt. We think of the rate as the derivative of concentration with respect to time. This derivative is the slope of a graph of concentration against time taken at a particular moment (*see* Fig. 5.1).

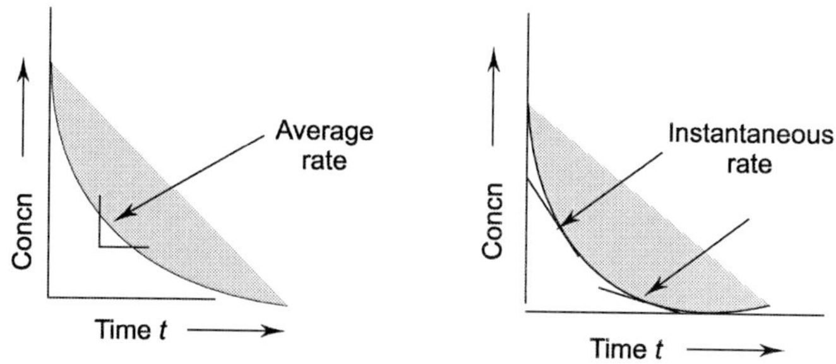

Fig. 5.1 Rates of reaction.

5.3 FACTORS INFLUENCING THE RATE OF REACTION

Experimental studies have shown that following factors influence the rate of a chemical reaction :

1. **Concentration of Reactants:** The Law of Mass Action clearly indicates that the rate of a chemical reaction is proportional to the product of active masses or molar concentrations of reacting species. The rate at which an iron piece is rusted is influenced by the presence of water vapour. Hence in rainy season, when water vapour concentration is tremendously high, the rate of rusting is also enhanced.
2. **Effect of Temperature:** Rate of reaction increases with rise of temperature. It has been experimentally found that for most of the reactions, the rate of reaction is doubled for every 10° C increase in temperature. In case of decomposition of hydrogen iodide, with an increase of 10°C temperature, the rate constant is increased by a factor of 1.7. This can be expressed as :

   Temperature coefficient = (rate constant at $(t + 10)$, *i.e.*, K_{t+10})/(rate constant at (t), *i.e.*, K_t)
   ≈ 2 to 3

   Generally, the temperature coefficient is taken as ratio of rate constants at 35°C and 25°C.

   $$\text{Temperature coefficient} = \frac{K_{35}}{K_{25}} \simeq 2$$

3. **Effect of Pressure:** Effect of pressure on the rate of reaction has been predicted by van't Hoff in terms of equilibrium constant. Consider the following reaction :

   $$A + BC \underset{K_{-1}}{\overset{K_1}{\rightleftharpoons}} AB + C$$

   where, K_1 and K_{-1} are the rate constants for forward and reverse reaction.

Variation of the equilibrium constant (K) for the reaction $\left(K = \frac{K_1}{K_{-1}}\right)$ with pressure is given by

$$\left(\frac{\delta \log K}{\delta P}\right)_T = -\left(\frac{\Delta V}{RT}\right)$$

Thus, when the reaction is accompanied by increase of volume (*i.e.*, ΔV is positive); the derivative of equilibrium constant with pressure decreases, *i.e.*, increasing pressure would decrease the proportions of product. On the other hand, when ΔV is negative the equilibrium constant increases with increasing pressure and hence the rate of product formation is also increased.

4. **Solvent:** The reaction between triethyl amine and ethyl iodide forming quaternary ammonium salt has been studied in 22 solvents and it has been concluded that the rate of reaction varied quite enormously. For example, rate constant at 100°C in hexane is 0.5×10^{-5}, while the same in bromobenzene and nitrobenzene is 166×10^{-5} and 1383×10^{-5}, respectively. Rate constant for thermal decomposition of trinitrobenzoic acid at 70°C is very less in toluene than it is in water. Probably the solvent in which the reactant gets solvated has pronounced influence on the rate of reaction. Very high rate of formation of quaternary ammonium salts in nitrobenzene as compared to benzene is explained on the basis of formation of solvated active complex in polar solvents. Ionic reactions in solutions are affected by electrostatic interactions.
5. **Light:** Photochemical reactions take place due to absorption of photons (energy of photon being given by $h\nu$, where ν is the frequency of light affecting the chemical reaction). Photochemical reaction between hydrogen and chlorine is started by the light absorbed by chlorine in the region of 4785 Å and rate of reaction is directly proportional to the first power of intensity of light absorbed.
6. **Catalyst:** Catalyst alters the rate of reaction without taking part in the reaction. The catalyst increases the rate of reaction by lowering the energy requirement of the rate determining step. It increases the rate of attainment of equilibrium without disturbing the equilibrium composition of the reaction system.

5.4 CONCENTRATION DEPENDENCE OF REACTION RATE

The Law of Mass Action says that the rate of a homogeneous reaction is proportional to some power of concentrations (mol l^{-1}) of the reactants. Thus, for a general reaction

$$aA + bB \rightarrow cC + dD$$

$$\text{Rate } \alpha \ C_A^x \ C_B^y$$

$$\text{Rate} = k \ C_A^x \ C_B^y$$

where, k is the proportionality constant. This way of mathematically presenting the rate of a chemical reaction is called **rate law**. The numerical value of x and y are to be determined experimentally and they could be different from stoichiometric coefficients a and b.

The proportionality constant k in equation is called **specific reaction rate** or **velocity constant** of the reaction. Unit of k would depend on the unit used for concentration and time.

Value of k varies with the temperature of reaction.

Sum of the powers of the concentration terms (x and y) appearing in the Rate Law or Rate Equation of all the concentration terms, *i.e.*, $(x + y)$ gives the overall order of reaction. However, the order with respect to a particular reactant, say A, is the value of exponent of the concentration term for A *i.e.*, x. Thus, the reaction is of order x with respect to A. Now we know that

$$\text{Rate} = \left(\frac{dx}{dt}\right) = -\left(\frac{1}{a}\right)\left(\frac{d\,C_A}{dt}\right) = -\left(\frac{1}{b}\right)\left(\frac{d\,C_B}{dt}\right)$$

$$\left(\frac{dx}{dt}\right) = k\,C_A^x\,C_B^y$$

This equation is referred as **differential rate law** or **differential rate equation** and correlates the rate of reaction with concentrations of all the reacting species.

Unit of k : When $(x + y) = n$, $k = \dfrac{\left(\frac{dx}{dt}\right)}{\left(C_A^x\,C_B^y\right)}$; let $(x + y) = n$

Thus unit for $K = (\text{conc.}) \times (\text{time})^{-1} / (\text{conc.})^n$

$= (\text{conc.})^{(1-n)} \times \text{time}^{-1}$

Thus when concentrations are expressed in mol l^{-1} and time in second (s)

Unit for $k = (\text{mol l}^{-1})^{(1-n)}\ \text{s}^{-1}$

where 'x' is the order of reaction.

5.5 MOLECULARITY AND ORDER OF REACTION

Molecularity of a reaction is the number of reacting molecules taking part (*e.g.*, colliding simultaneously) in an elementary step of the reaction. Reaction like

$$\text{Hexene} \xrightarrow{\Delta} \text{Cyclohexane}$$

is a unimolecular reaction as only one molecule is taking part in the reaction. Similarly bimolecular and termolecular reactions are :

$$\text{Bimolecular}\ \ 2H_2O_2 \longrightarrow 2H_2O + O_2$$

$$\text{Termolecular}\ \ 2NO + O_2 \longrightarrow 2NO_2$$

Generally, reactions involving three or more molecules in an elementary step are rare in occurrence as the simultaneous collision of such a large number of molecules is a rare event. Such reactions do not occur in one step but in a sequence of consecutive steps and are called complex reactions. As a result molecularity of such reaction has no significance as far as rate of reaction is concerned.

Order of reaction on the other hand depends on the number of molecules whose change in concentration alters the rate of reaction. Thus, for the reaction

$$xA + yB + zC \longrightarrow \text{Products}$$

the order of reaction depends on the rate equation.

If the rate equation is

$$\text{rate} = k\,[A]^l\,[B]^m\,[n]^n$$

then order of reaction is $[l + m + n]$ and order with respect to $[A]$ is $[l]$, with respect to $[B]$ is $[m]$ and with respect to $[C]$ is $[n]$.

Similarly if rate of above equation is rate $= k\,[A]^P\,[B]^q$ then overall order of reaction is $[p + q]$ and order with respect to $[A]$ and $[B]$ is $[p]$ and $[q]$, respectively.

If rate of above equation is rate $= k\,[A]^z$ then order of reaction is $[z]$ and with respect to $[A]$ reaction is of $[z]$ order. It is not necessary that l, m, n, p, q, z, *etc.*, are whole numbers and hence order can be fractional also.

Gaseous reactions involving decomposition of gases on catalyst surface are generally unimolecular. If the gas is strongly absorbed at the catalyst surface the rate becomes independent of the concentration of gas (as it now depends on available catalyst surface) and the reaction becomes zero order. However, if the gas is weakly adsorbed on the catalyst surface, the rate depends on the concentration or pressure of the gas and the reaction becomes first order. In all intermediate cases the order will be fractional.

There are a number of examples where a bimolecular reaction is found to be first order, *e.g.*,

$$2N_2O_5 \longrightarrow 4NO_2 + O_2$$

$$2H_2O_2 \longrightarrow 2H_2O + O_2$$

or it may be pseudo first-order, *e.g.*,

$$CH_3COOCH_3 + H_2O \xrightarrow{H^+} CH_3COOH + CH_3OH$$

$$\underset{\text{sucrose}}{C_{12}H_{22}O_{11}} + H_2O \xrightarrow{H^{\oplus}} \underset{\text{glucose}}{C_6H_{12}O_6} + \underset{\text{fructose}}{C_6H_{12}O_6}$$

Decomposition of acetaldehyde into methane and carbon monoxide is an example of fractional order where order has been observed to be 1.5.

5.6 MATHEMATICAL CHARACTERISTICS OF SIMPLE REACTIONS

5.6.1 Zero Order Reactions

Suppose the initial concentration of the reactant is a, the rate of a zero order reaction would be

$$\left(\frac{dx}{dt}\right) \alpha\, a^0$$

or

$$\left(\frac{dx}{dt}\right) = k$$

Here k is the specific reaction rate constant of zero order reaction. Separating the variables, we get

$$dx = k\,dt$$

On integration, we get

$$x = kt + I$$

where, I, the integration constant can be evaluated by putting $x = 0$ at $t = 0$, and thus $I = 0$, hence

$$x = kt \tag{5.1}$$

Half–life period ($t_{1/2}$) is the time taken for half of the reaction to complete and this may be calculated as : when $x = \frac{a}{2}$ (half of the initial concentration of the reactant is consumed), $t = (t_{1/2})$,

or $$\frac{a}{2} = k\,(t_{1/2}),\ \text{so}\ (t_{1/2}) = \frac{a}{(2k)}$$

Thus, half-life period of zero order reactions is proportional to the first power of initial concentration.

From Eq. 5.1 it can be concluded that

1. Plot of x vs time would be a straight line passing through origin. Thus, the concentration of the product increases with time in a linear fashion.
2. Unit of rate constant for zero order reaction is mol l^{-1} s^{-1}.

■ **Example 1.** Decomposition of ammonia.

A study of the decomposition of ammonia into nitrogen and hydrogen on tungsten and platinum surface has been undertaken

$$2\ NH_3 \longrightarrow N_2 + 3\ H_2$$

Ammonia gas at a known pressure is taken in tungsten–filament lamp. It is cooled to ice temperature and then lamp is heated electrically to a constant temperature of 850°C. Pressure of the decomposed gases is measured manometrically and plotted vs time. It has been found that the initial slopes with different initial concentrations remain the same. This indicated that the reaction is of zero order.

■ **Example 2.** Decomposition of N_2O on a hot platinum surface is a zero order reaction and can be shown as:

$$N_2O \longrightarrow N_2 + \left(\frac{1}{2}\right) O_2$$

$$\text{Rate}\ \alpha\ C^0_{N_2O}$$

$$\text{Rate} = -\left(\frac{d\,C_{N_2O}}{dt}\right) = k\,C^0_{N_2O} = k = \text{constant}$$

Problem 1. The initial concentration of a zero order reaction is 0.15 mol l^{-1} and after half an hour its concentration goes to 0.05 mol l^{-1}. Find its rate constant.

Solution: For a zero order reaction,

$$k = \frac{x}{t} = \frac{(0.15 - 0.05)}{30 \times 60} = \frac{0.1}{1800} = 0.55 \times 10^{-4} \text{ mol } l^{-1} \text{ s}^{-1}.$$

5.6.2 First Order Reactions

When the sum of the exponents of the concentration terms appearing in the rate law is one, it is a reaction of the first order and it would be represented as :

$$A \longrightarrow \text{Products}$$

Let us suppose that the initial concentration of a reactant A is a mol l^{-1} and x mol l^{-1} of it is decomposed in time t. Hence, rate of reaction at time t is given by

$$\frac{dx}{dt} = k\,(a - x)$$

where, k is the specific reaction rate or velocity constant. This equation can be represented as:

$$\frac{dx}{(a - x)} = k\,dt$$

This on integration gives

$$\int \frac{dx}{(a - x)} = k \int dt$$

$$-\log_e (a - x) = kt + I$$

I is the constant of integration and is evaluated by putting $x = 0$ at $t = 0$.

Thus $$I = -\log_e a,$$

hence $$k = \left(\frac{1}{t}\right) \log_e \frac{a}{(a - x)}$$

or $$k = 2.303 \left(\frac{1}{t}\right) \log_{10} \frac{a}{(a - x)} \qquad (5.2)$$

This is the kinetic equation of first order reactions.

Characteristics of the First Order Reactions

1. Concentrations of the reactants can be determined at different time intervals and substituted in Eq. 5.2. If the values of k remain constant for different values of initial concentrations, the reaction follows the first order equation.

2. **Half-Life Period for the First Order Reactions:** Suppose in time, $t_{1/2}$ half of the initial concentration of a reactant is reacted, *i.e.*, $x = \frac{a}{2}$, then

$$t_{1/2} = \left(\frac{2.303}{k}\right) \log_{10} \left[\frac{a}{\left(a - \frac{a}{2}\right)}\right]$$

$$t_{1/2} = \left(\frac{2.303}{k}\right) \log_{10} 2 = \frac{0.693}{k}$$

3. **Units of First Order Rate Constant:**

$$k = \left(\frac{1}{\text{time}}\right)\left[\frac{(\text{concentration})}{(\text{concentration})}\right]$$

$$K = (\text{time})^{-1} \text{ or } s^{-1}.$$

4. **Physical Significance of *k*:**

Equation 5.2 may be arranged as

$$\log_{10}(a - x) = \left(-\frac{k}{2.303}\right) t + \log_{10} a$$

This is a equation of straight line. Hence, if a plot is made between values of $\log_{10}$ $(a - x)$ vs time (t), Fig. 5.2 shall be obtained and value of specific reaction rate constant k, can be derived from its

$$\text{Slope} = \left(-\frac{k}{2.303}\right)$$

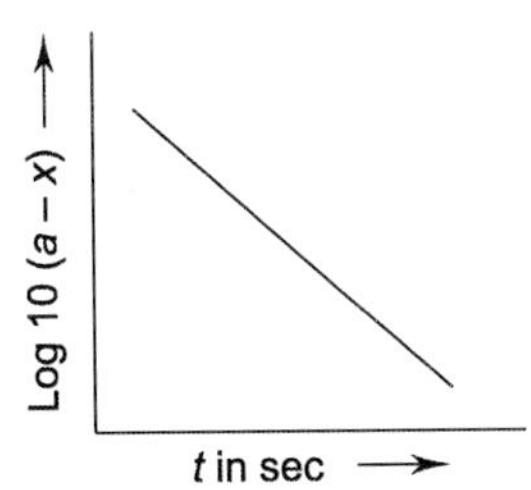

Fig. 5.2

■ **Example 1.** Thermal decomposition of nitrogen pentoxide.

The decomposition of nitrogen pentoxide is the first example of homogeneous reaction to be investigated in this class

$$2N_2O_5 \rightleftharpoons 2N_2O_4 + O_2$$

The rate of this reaction can be expressed as

$$\left(-\frac{dC_{N_2O_5}}{dt}\right) = kC_{N_2O_5}$$

Order of the reaction has been found to be one. Oxygen pressure is measured at different time intervals. Suppose V_t is the volume of gas measured at time t and V_∞ is the final volume when reaction is complete. Then $a \alpha V_\infty$ and $(a-x) \alpha (V_\infty - V_t)$; thus

$$K = \left(\frac{2.303}{t}\right) \log_{10}\left[\frac{a}{(a-x)}\right]$$

$$= \left(\frac{2.303}{t}\right) \log_{10} \frac{V_\infty}{(V_\infty - V_t)}$$

■ **Example 2.** Decomposition of hydrogen peroxide.

The decomposition of H_2O_2 into water and oxygen is a first order reaction :

$$2H_2O_2 \longrightarrow 2H_2O + O_2$$

The kinetics can be studied by estimating the concentration of H_2O_2 by titrating it against standard $KMnO_4$.

If the initial reading of $KMnO_4$ is V_0 and reading at time t is V_t, then value of rate constant

$$K = \frac{2.303}{t} \log \frac{V_0}{V_t}.$$

Problem 2. From the following data, show that the decomposition of hydrogen peroxide in aqueous solution is a first order reaction.

Time in Seconds	0	300	600	1200
Volume of $KMnO_4$ used to titrate H_2O_2.	46.0	37.0	30.0	20.0

Solution: For a first order reaction $K = \dfrac{2.303}{t} \log \dfrac{a}{a-x}$

Here $a \propto V_0$ and $a - x \propto V_t$. Substituting these values in above equation for

$$t = 300 \text{ s} \quad K_1 = \frac{2.303}{300} \log \frac{46.0}{37.0} = 0.00072 \text{ s}^{-1}$$

$$t = 600 \text{ s} \quad K_1 = \frac{2.303}{600} \log \frac{46.0}{30.0} = 0.0007 \text{ s}^{-1}$$

$$t = 1200 \text{ s} \quad K_1 = \frac{2.303}{1200} \log \frac{46.0}{20.0} = 0.0007 \text{ s}^{-1}.$$

The constancy of K (velocity constant) show that it is a first order reaction.

5.6.3 Pseudo Unimolecular Reactions : Pseudo Order Reactions

Order of the reaction is an experimentally determined quantity. In certain reactions it is found that 'a' moles of A and 'b' moles of B are participating and thus order of reaction, apparently may be given by $(a + b)$. When experimental obervations suggest that the rate of reaction is dependent only on the concentration of say A only, then it is a case of **pseudo order**, and the actual order of the reaction will be only 'a' and not $(a + b)$. Examples of pseudo first order reaction are given below :

■ **Example 1.** Hydrolysis of methyl acetate

$$CH_3COOCH_3 + H_2O \xrightarrow{H^{\oplus}} CH_3COOH + CH_3OH$$

Although two molecules, *i.e.*, one molecule of methyl acetate and another of water are involved in this reaction, experimentally it has been found that the rate is determined by the concentration of methyl acetate only and not by concentration of water. Small changes in the concentration of water are negligible compared to the large excess of water present.

The progress of the reaction is measured by titrating the acetic acid formed. A definite volume of reaction mixture is withdrawn at different intervals of time and titrated against standard alkali (NaOH) solution. Final reading is taken after 24 h and the value of k can be calculated from

$$K \equiv \left(\frac{2.303}{t}\right)\log_{10}\left[\frac{(V_\infty - V_0)}{(V_\infty - V_t)}\right]$$

where,

$V_0 \equiv$ Titer value at the beginning of the reaction

$V_t \equiv$ Titer value at the time t

$V_\infty \equiv$ Titer value after 24 h.

■ **Example 2.** Inversion of cane sugar

The hydrolysis reaction of sucrose (cane sugar) is

$$\underset{\text{sucrose}}{C_{12}H_{22}O_{11}} + H_2O \xrightarrow{H^{\oplus}} \underset{\text{glucose}}{C_6H_{12}O_6} + \underset{\text{fructose}}{C_6H_{12}O_6}$$

Although two molecules are involved in the reaction, experimentally it has been found that the rate of hydrolysis is dependent only on the sucrose concentration. Hence it is also an example of pseudo first–order reaction. The progress of this reaction is followed by measurement of angle of optical rotation of reaction mixture. Although sucrose rotates the plane of polarization of light to the right, the mixture obtained after hydrolysis turns it to the left. The change produced in the rotation in the time t gives the value of x. Suppose the optical rotation of sucrose solution at the beginning of the reaction is r_0 and is r_t after time t and r_∞ at the end of the reaction. Then,

$$a \equiv (r_0 - r_\infty) \text{ and } x \equiv (r_0 - r_t)\ ;\ (a - x) \equiv (r_t - r_\infty)$$

$$K = \left(\frac{2.303}{t}\right)\log_{10}\left[\frac{a}{(a-x)}\right]$$

$$K = \left(\frac{2.303}{t}\right)\log_{10}\left[\frac{(r_0 - r_\infty)}{(r_t - r_\infty)}\right]$$

Problem 3. Optical rotations of cane sugar in 0.5 N HCl at 35°C and at various intervals is given below. Show that the reaction is of first order :

Time (min.)	0	20	40	80	∞
Rotation (degrees) optical	+ 34.0	+ 20.6	+ 15.6	+ 11.3	– 10.74

Solution : Here $r_0 - r_\infty \propto a$ and $r_t - r_\infty \propto (a - x)$

$$(r_0 - r_\infty) = [+34.0 - (-10.74)] = +44.74 = a$$

At $t = 20$, $(r_t - r_\infty) = [+25.5 - (-10.74)] = +36.24 = (a - x)$

$$K = \left(\frac{2.303}{20}\right)\left(\log_{10}\frac{44.74}{31.34}\right) = 0.015 \text{ min}^{-1}$$

At $t = 40$, $(r_t - r_\infty) = [+15.6 - (-10.74)] = 26.34$

$$K = \left(\frac{2.302}{40}\right)\left(\log_{10}\frac{44.74}{26.34}\right) = 0.013 \text{ min}^{-1}.$$

At $t = 80$, $(r_t - r_\infty) = [+11.3 - (-10.74)] = +22.04$

$$k = \left(\frac{2.303}{80}\right)\left(\log_{10}\frac{44.74}{22.04}\right) = 0.011 \text{ min}^{-1}.$$

As the value of k being calculated from first order equation is almost constant, it is concluded that inversion of sucrose in an example of a first order reaction.

5.6.4 Second Order Reactions

Case I: When two molecules of the same reactants are reacting, we can write

$$2A \longrightarrow \text{Products}$$

If the initial concentration of the reactant A is a mol l^{-1} and after time t, mol l^{-1} of it have reacted, then concentration of A remaining will be $(a - x)$ mol l^{-1}. Rate of reaction for a second order reaction is proportional to the second power of concentration of the reactant, we get

$$\frac{dx}{dt} = k\,(a - x)^2$$

where, k is the rate constant. This on rearrangement and implementation of integration sign gives

$$\int \frac{dx}{(a - x)^2} = k \int dt$$

On integrating this equation

$$\frac{1}{(a-x)} = kt + I, \text{ where } I \text{ is integration constant.}$$

Implementing the condition that, when ($x = 0$, $t = 0$) ; therefore, $I = \frac{1}{a}$

$$kt = \frac{1}{(a-x)} - \frac{1}{a} = \frac{x}{a(a-x)}$$

or

$$k = \left(\frac{1}{t}\right)\frac{x}{a}\frac{1}{(a-x)} \tag{5.3}$$

This is the integrated rate equation for a second order reaction.

Case II: When both the reactants are different and each reactant is in different initial concentration, we can then write the reaction as

$$A + B \longrightarrow \text{Products}$$

Rate of the reaction is then expressed as

$$\text{Rate} = -\left(\frac{dC_A}{dt}\right) = -\left(\frac{dC_B}{dt}\right) = k\,C_A\,C_B$$

It is assumed that initial concentration of A is a mol l^{-1} and that *of* B is b mol l^{-1}, and during time t, x mol l^{-1} of A have reacted with x mol l^{-1} of B and have formed x mol l^{-1} of the products. Thus, concentrations of A and B present at time t is $(a - x)$ and $(b - x)$ mol l^{-1}, respectively. So, we can express the rate as

$$\text{Rate} = -\left(\frac{dC_A}{dt}\right) = \frac{dx}{dt} = k(a-x)(b-x)$$

where, k is the second order rate constant. Equation can be solved by separating the variables and putting as

$$\frac{dx}{(a-x)(b-x)} = k\,dt$$

Resolving the fraction of left hand side into partial fractions, we get

$$\frac{1}{(a-x)(b-x)} = \frac{1}{(a-b)}\left[\frac{1}{(b-x)} - \frac{1}{(a-x)}\right]$$

Hence, integration on both the sides

$$\int\frac{dx}{(a-x)(b-x)} = \frac{1}{(a-b)}\left[\int\frac{dx}{(b-x)} - \int\frac{dx}{(a-x)}\right] = k\,dt$$

We get $$\frac{1}{(a-b)}\left[-\log_e(b-x)-\{-\log_e(a-x)\}\right]=kt+I \qquad (5.4)$$

where, I is the integration constant, and can be determined by applying the condition that when $t = 0$, $x = 0$, hence

$$I=\frac{1}{(a-b)}\log_e\left(\frac{a}{b}\right)$$

Substituting this value of I in Eq. 5.4, we get

$$\frac{1}{(a-b)}\log_e\frac{(b-x)}{(a-x)}=kt+\frac{1}{(a-b)}\log_e\left(\frac{a}{b}\right)$$

or $$k=\left(\frac{1}{t}\right)\frac{2.303}{(a-b)}\log_{10}\left[\frac{b(a-x)}{a(b-x)}\right] \qquad (5.5)$$

This is the integrated equation of the second order reaction with different reactants. Putting this equation as

$$\frac{2.303}{(a-b)}\log_{10}\left[\frac{b(a-x)}{a(b-x)}\right]=kt$$

This is an equation for a straight line ($y = mx + c$). Hence, plotting the values of $\log_{10}\left[\frac{b(a-x)}{a(b-x)}\right]$ vs t should furnish a straight line Fig. 5.3. The slope of this line is $\frac{2.303}{k(a-b)}$ and thus

$$k=\left[\frac{2.303}{(a-b)}\right]\times\frac{1}{\text{slope}}$$

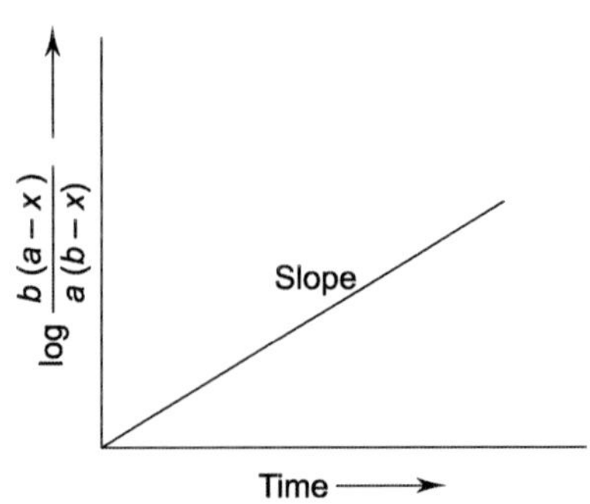

Fig. 5.3

Characteristics of Second Order Reactions:

1. From Eq. 5.3, $k = \left(\frac{1}{t}\right)\frac{x}{a}(a-x)$, we can derive the units for second order velocity constant,

$$k = \left(\frac{1}{\mathrm{s}}\right)\left[\frac{(\mathrm{mol\,l^{-1}})}{(\mathrm{mol\,l^{-1}})(\mathrm{mol\,l^{-1}})}\right]$$

$$k = \mathrm{l\ mol^{-1}\ s^{-1}}$$

2. Time for half-change : When $x = \frac{a}{2}$; $t = t_{1/2}$, hence from Eq. 5.3

$$t_{1/2} = \left(\frac{1}{k}\right)\left[\frac{(a/2)}{a(a-a/2)}\right] = \frac{1}{ka}$$

So, in second order reactions time required for half of the reaction to complete is inversely proportional to initial concentration.

■ **Example 1.** Saponification of ethyl acetate : the reaction between ethyl acetate and sodium hydroxide takes pl\ace as

$$CH_3COOC_2H_5 + NaOH \longrightarrow CH_3COONa + C_2H_5OH$$

50 mL solutions of ethyl acetate (0.01*M*) and sodium hydroxide (0.2*M*) are taken separately in two flasks and kept in a thermostatic bath to attain the desired temperature. The reaction is started as soon as the alkali solution is added to ester and shaken. 5 mL of the reaction mixture is taken out, through a pipette, in another flask in which 10 mL of 0.04*M* HCl has already been taken. This will stop the continuation of reaction and excess of acid could be back titrated using standard alkali solution. Let the titre value be V_0. Similarly the reaction mixture (5 mL) would again be withdrawn regularly after 10, 20, 30 and 40 min and added to 10 mL of 0.04*M* acid to check the continuation of the reaction and excess acid back titrated to furnish V_t values. The V_∞ reading is then obtained by continuing the reaction for 24 hours and then following the same procedure. $V_0 \propto b$, $(V_0 - V_t) \propto x$, $(V_0 - V_\infty) \propto a$. Substitution of these in Eq. 5.5 would give a constant value of specific reaction rate (k) for the saponification of ethyl acetate, thereby indicating that the reaction follows second order kinetics.

Problem 4. Hydrolysis of ethyl acetate using equal concentrations of ester and NaOH was studied by titrating the reaction mixture with standard HCl at different time intervals when following results were obtained

Time (s)	0	225	675	1125
Volume of 0.1 N HCl used on mL	24.0	14.0	7.5	5.2

Solution : Equation of second order when the concentration of both reactant is same.

$$K = \frac{1}{t}\frac{x}{a(a-x)}$$

Here $a \propto NV_0 = 0.1 \times 24 = 2.4$

At time $t\ (a - x) \propto NV_t$

Hence $x \propto N\,(V_0 - V_t)$

For $t = 225$ s, $a = 2.4$, $a - x = 0.1\ (14) = 1.4$ and $x = 2.4 - 1.4 = 1.0$

$$K_1 = \frac{1}{225} \times \frac{1.0}{2.4 \times 1.4} = 1.3 \times 10^{-3}\ \text{mol}^{-1}\ \text{l s}^{-1}$$

For $t = 675$ s, $a = 2.4$, $a - x = 0.1\ (7.5) = 0.75$ and $x = 2.4 - 0.75 = 1.65$

$$K_2 = \frac{1}{675} \times \frac{1.65}{2.4 \times 0.75} = 1.3 \times 10^{-3}\ \text{mol}^{-1}\ \text{l s}^{-1}$$

For $t = 1125$ s, $a = 2.4$, $a - x = 0.1\ (5.2) = 0.52$ and $x = 2.4 - 0.52 = 1.88$

$$K_3 = \frac{1}{1125} \times \frac{1.88}{2.4 \times 0.52} = 1.3 \times 10^{-3}\ \text{mol}^{-1}\ \text{l s}^{-1}.$$

Since the values of velocity constants K_1, K_2 and K_3 are fairly constant when experimental data is substituted in second order equation, the reaction is of second order.

5.6.5 Third Order Reaction

Consider a hypothetical reaction

$A + B + C \longrightarrow$ products where rate of reaction is given by

$$-\frac{dA}{dt} = -\frac{dB}{dt} = -\frac{dC}{dt} = k\,[A]\,[B]\,[C]$$

If initial concentration of A, B and C are a, b and c, respectively, and decrease in their concentrations after time t is x, then rate equation becomes

$$\frac{dx}{dt} = k\,(a - x)\,(b - x)\,(c - x)$$

In case $a = b = c$ or for a third order reaction of the type $3A \rightarrow$ products, rate equation becomes

$\frac{dx}{dt} = k\,(a - x)^3$ which on integration yields equation for third order,

$$K_1 = \frac{1}{2t}\left[\frac{1}{(a-x)^2} - \frac{1}{a^2}\right]$$

The half-life period for such a reaction is given by

$$t_{1/2} = \frac{3}{2K_1\, a^2}$$

Examples of third order reaction are few. Some of these are

$$2NO + X_2 \rightarrow \text{product}$$

where X_2 is a halogen Cl_2 or Br_2

and $$SnCl_2 + 2FeCl_3 \rightarrow 2FeCl_2 + SnCl_4$$

Reactions of yet higher order are unknown.

5.7 RADIOACTIVE DECAY AS A FIRST ORDER PHENOMENON

The radioactive disintegrations of all radioactive substances are basically of first order nature. When an alpha particle $\left({}_2^4\text{He}\right)$ is emitted, the resulting nucleus (daughter elements) has a mass less by four units and charge less by two units as compared with parent nucleus, *e.g.*,

$${}^{226}_{88}\text{Ra} \rightarrow {}^{222}_{86}\text{Rn} + {}^{4}_{2}\text{He}$$

In case a β-particle ($-1e^{\circ}$) is emitted the resulting daughter nucleus has the same mass but nuclear charge increases by one unit.

The activity of a radioactive element is measured, using GM Counter, by the rate at which it changes into its daughter element. According to the Law of Radioactive Decay, the quantity of a radioactive element which disappears in unit time (rate of disintegration) is directly proportional to the amount present. The radioactive decay follows first order kinetics as has been shown below :

Suppose initially N atoms of A are present in a decay given below

$$A \rightarrow B$$

If the number of atoms decaying in time, dt, is dN, then rate of disintegration is given as

$$-\left(\frac{dN}{dt}\right) \propto N$$

or $$-\left(\frac{dN}{dt}\right) = \lambda N$$

where λ is the proportionality constant and better known as **radioactive or disintegration or decay constant.** This equation may now be written as

$$-\frac{dN}{N} = \lambda\, dt$$

Integrating this equation, we get

$$-\int \frac{dN}{N} = \lambda \int dt \quad \text{or} \quad -\log_e N = \lambda t + I \tag{5.6}$$

Here I is the constant of integration. We know that when $t = 0$, the number of disintegrating atoms present is N_0. Hence, from Eq. 5.6, we get

$$-\log_e N_0 = 0 + I$$

or

$$-\log_e N = \lambda t - \log_e N_0$$

or

$$\log_e \frac{N_0}{N} = \lambda t$$

Thus,

$$\lambda = \left(\frac{2.303}{t}\right) \log_{10} \frac{N_0}{N} \tag{5.7}$$

This is the integrated equation expressing decay constant and number of disintegrating atoms present and is identical to first order equation.

Half-life Period ($t_{1/2}$): It is the time during which one-half of any sample disintegrates. Thus, if N_0 is the number of atoms present in the initial sample of a radioactive substance, then number of atoms present after $t_{1/2}$ will be $\frac{N_0}{2}$ and $\frac{N_0}{N_0/2} = 2$

Hence from Eq. 5.7,

$$\lambda = \left(\frac{2.303}{t_{1/2}}\right) \log_{10} \frac{N_0}{N} = \left(\frac{2.303}{t_{1/2}}\right) \log_{10} 2$$

or

$$t_{1/2} = \left(\frac{2.303}{\lambda}\right) 0.3010 = \frac{0.693}{\lambda} \tag{5.8}$$

Thus half-life period of a radioactive substance is independent of the amount of material present. Half-life for $^{222}_{86}\text{Rn}$ is 382 days, while it is 1.62×10^2 years for $^{226}_{88}\text{Ra}$.

Average-life Period : Total decay period of a radioactive element is usually very long as disintegration process may even continue for infinite period. Hence, another useful term, Average-life period (t_{av}) – which is reciprocal of its disintegration constant (λ) is more often utilized. Thus,

$$\lambda = \frac{0.693}{t_{1/2}}$$

and average-life period

$$(t_{av}) = \frac{1}{\lambda} = \frac{\left(t\sqrt{\tfrac{1}{2}}\right)}{0.693} = 1.44\left(t\sqrt{\tfrac{1}{2}}\right) \tag{5.9}$$

Problem 5. For a certain reaction it takes 5 min for the initial concentration of 0.5 mol l^{-1} to become 0.25 mol l^{-1} and another 5 min to become 0.125 mol l^{-1}. What is the specific reaction rate of this reaction ?

Solution : Since the half-life period is independent of the initial concentration of the reactant, the reaction is of first order. Hence,

$$t_{1/2} = \frac{0.693}{k}$$

or $$k = \frac{0.693}{t_{1/2}} = \frac{0.693}{5}$$

$$k = 0.139 \text{ min}^{-1}$$

Problem 6. The half-life period for the decomposition of radium (Ra) is 1.62×10^3 years. Calculate the rate constant in s^{-1}. In how many years will three quarters of a given amount of radium disappear ?

Solution : $t_{1/2}$ for radium = 1.62×10^3 years,

Now $$t_{1/2} = \frac{0.693}{\lambda} \text{ or } 1.62 \times 10^3 = \frac{0.693}{\lambda}$$

or $$\lambda = \frac{0.693}{1.62 \times 10^3} \text{ year}^{-1} = 1.36 \times 10^{-11} \text{ s}^{-1}$$

Now, it is given that when $\frac{3}{4}$ amount is consumed, the fraction remaining = $\frac{1}{4}$ or $\frac{N}{N_0} = \frac{1}{4}$ or $\frac{N_0}{N} = 4$.

$$\lambda = \left(\frac{2.303}{t}\right) \log_{10} \frac{N_0}{N} = \left(\frac{2.303}{t}\right) \log_{10} 4$$

So, $$t = 2.303 \times \frac{0.6021}{\lambda} = \frac{(2.303 \times 0.6021)}{1.36 \times 10^{-11}} \text{ s}^{-1}.$$

Or $t = 3240$ years ; thus three quarters of given radium would disappear in 3240 years or in two half-lives.

5.7.1 Radio-Dating

Isotopes of some radioactive substances having long half-lives may be used to determine the age of a sample of rock or of a fossil.

5.7.2 Mineral Dating

The age of a rock or meteorite sample may be determined by following the decay of ^{238}U to ^{206}Pb or ^{4}He, ^{232}Th to ^{208}Pb.

In case of ^{238}U decay, it is assumed that all the ^{206}Pb present in the mineral has come from the decay of ^{238}U. Knowing the half–life of ^{238}U the age of mineral can be determined as follows :

Problem 7. In a sample of mineral obtained from a meteorite it was observed that ^{238}U and ^{206}Pb are present in the atomic ratio of 1 : 0.1.

Given that half-life of ^{238}U is 4.5×10^9 years, calculate the approximate age of mineral.

Solution : It may be assumed that all the ^{206}Pb has come from the decay of ^{238}U. Hence, initial atoms of

$$^{238}U = 1 + 0.1 = 1.1\ (N_0)$$

Atoms of ^{238}U now $= 1.0\ (N_t)$

For decay of ^{238}U $k = \frac{0.693}{4.5 \times 10^9}\ year^{-1}$

$$k = \frac{2.303}{t} \log \frac{N_0}{N_t} \quad \text{or } t = \frac{2.303}{k} \log \frac{N_0}{N}$$

$$\therefore \quad t = \frac{2.303}{0.693} \times 4.5 \times 10^9 \log \frac{1.1}{1.0} = 6.19 \times 10^8 \text{ years.}$$

5.7.3 Fossil Dating or Carbon Dating

W.F. Libby observed that in the natural atmosphere the ratio of ^{14}C and ^{12}C has remained constant for last several thousand years. Living systems–plants or animals continuously exchange CO_2 with environment. However, the moment plant or animal dies this exchange of CO_2 with atmosphere ceases. The ^{14}C trapped in the body at the time of death starts decaying.

Thus by determining ^{14}C activity in a fossil sample and comparing it with ^{14}C activity in atmosphere the age of a fossil sample can be determined.

Problem 8. A sample of wooden fossil is found to emit 7.4 counts per minute per gram of carbon. A fresh sample of wood emits 15.3 counts per minute per gram of carbon.

If the half–life of ^{14}C is 5660 years, find approximate age of sample.

Solution : The decay of ^{14}C follows first order kinetics

$$k = \frac{2.303}{t} \log \frac{a_0}{a_t} \quad \text{and } k = \frac{0.693}{5660}\ \text{year}^{-1}$$

$$a_0 = 15.3 \text{ c.p.m. and } a_t = 7.4 \text{ c.p.m.}$$

$$\therefore \quad t = \frac{2.303}{0.693} \times 5660 \log \frac{15.3}{7.4} = 5936 \text{ years.}$$

Hence, the sample is 5936 years old.

5.8 DETERMINATION OF ORDER OF REACTION

5.8.1 Differential Method

Taking care of van't Hoff interpretation that rate of reaction of nth order should be proportional to the nth power of concentration of the reactant, we may write

$$-\frac{dc}{dt} \alpha C^n \text{ or } -\frac{dc}{dt} = k C^n$$

where, C is the concentration of the reactant. Taking logarithm of equation

$$\log_e \left(-\frac{dc}{dt}\right) = \log_e k + n \log_e C$$

$$\log_{10} \left(-\frac{dc}{dt}\right) = -\log_{10} k + n \log_{10} C \quad (5.10)$$

First we plot concentration vs time curve based on experimental data and obtain $\left(-\frac{dc}{dt}\right)$ from the slope of the curve at a particular instant. Then a straight line plot of $\log_{10}\left(-\frac{dc}{dt}\right)$ vs $\log_{10} C$ is obtained. The slope of this line gives order (n) and its intercept on y-axis furnishes k, the velocity constant of the reaction. A particular reaction may be carried out at different initial concentrations, say C_1 and C_2, and its rate $\left(-\frac{dc}{dt}\right)$ may be measured experimentally. Then we have

$$\left[\log_e\left(-\frac{dc_1}{dt}\right)\right] - \left[\log_e\left(-\frac{dc_2}{dt}\right)\right] = n\,[\log_e C_1 - \log_e C_2]$$

or

$$n = \left\{\frac{\left[\log_e\left(-\frac{dc_1}{dt}\right)\right] - \left[\log_e\left(-\frac{dc_2}{dt}\right)\right]}{\left[\log_e C_1 - \log_e C_2\right]}\right\} \quad (5.11)$$

This then leads to calculation of order of reaction n.

5.8.2 Integration Method

This is the most common method for determining order of simple reactions. Change of concentration vs time data is substituted in the rate equations of 1st, 2nd, 3rd order reactions and the values of specific reaction rate, k, are calculated. The order of reaction corresponds to that equation for which the various values of k obtained for different time intervals are constant.

5.8.3 Half-Life Method

When all the reactants are present in same molar concentrations, the half-life time ($t_{1/2}$) is related to the initial concentration of the reactant in a reaction of nth order as

$$t_{1/2} \propto \left(\frac{1}{a^{n-1}}\right)$$

Thus, for a reaction performed with two different initial concentrations

$$\frac{(t_{1/2})_1}{(t_{1/2})_2} = \left(\frac{a_2}{a_1}\right)^{n-1} \qquad (5.12)$$

Taking logarithm

$$\log_e\left[\frac{(t_{1/2})_1}{(t_{1/2})_2}\right] = (n-1)\log_e\left(\frac{a_2}{a_1}\right)$$

or

$$n = 1 + \frac{\log_e\left[\frac{(t_{1/2})_1}{(t_{1/2})_2}\right]}{\log_e\left(\frac{a_2}{a_1}\right)} \qquad (5.13)$$

Thus, knowing half-life times for the same reaction carried out at two different initial concentrations, the order of reaction (n) can be computed from this equation.

5.8.4 Isolation Method

We have seen in pseudo-order discussion that when a certain reactant is taken in excess compared to other reactants, the one in excess does not influence the order of the reaction. So in a reaction in which more than one reactants are present, we can isolate one among several reactants by taking all except one in excess. Here we get order of reaction with respect to that reactant which has been isolated. Next time we carry out experiment isolating another reactant, and so on. Hence the overall order of reaction will be the sum of the orders obtained by isolating the reactants one by one.

Problem 9. The time for half-change of a reaction undergoing catalytic thermal decomposition was measured for various initial pressures (P), and gave the following observation

P (cm)	10	20	30
t (s)	3.50	2.0	1.0

Find out the order of reaction.

Solution : We know that for this reaction

$$\frac{(t_{1/2})_1}{(t_{1/2})_2} = \left(\frac{a_2}{a_1}\right)^{n-1} = \left(\frac{p_2}{p_1}\right)^{n-1}$$

When $p_1 = 10$ cm; $(t_{1/2})_1 = 3.50$ s

$p_2 = 20$ cm; $(t_{12})_2 = 2.0$ s

So $$\frac{3.50}{2.0} = \left(\frac{20}{10}\right)^{n-1}$$

or $$\log_{10}\left(\frac{3.50}{2.0}\right) = (n-1)\log_{10} 2$$

thus $$n = 1.8 \approx 2$$

5.9 EXPERIMENTAL METHODS OF CHEMICAL KINETICS

In studying chemical kinetics, our goal is to establish a satisfactory mechanism of the chemical reaction. In this regard the first step is to know the order of reaction. This requires the determination of concentration of various reacting and product species with respect to time. An analytical method which is selected for measuring the concentration of a chemical reactant or product must be the one through which the reactants/ products are satisfactorily analysed and other reactants present are not interfering in this procedure. For example, in the hydrolysis of cane sugar into glucose and fructose, the kinetic studies are made by measuring the optical rotation of the reaction mixture with the help of polarimeter. This is possible due to the fact that during the course of reaction the optical rotation of the reaction mixture undergoes a remarkable change.

Similarly in most of the reactions studied in solution, absorption spectrophotometry is commonly employed to determine the concentration of reactants or products. The characteristic absorption band in the UV/visible absorption spectrum of either reactant or product undergoes a change in absorbance as a function of time. Some of these methods are described below :

5.9.1 Conductometric Titration

Conductivity measurement is a very convenient analytical tool for studying kinetics of reaction, involving formation and removal of ions in the reaction mixture. The conductance of a solution can be correlated with concentrations of various ions in solution (C_i) as

$$\text{Conductance} = \frac{1}{R} = x\left(C_1\lambda_1 + C_2\lambda_2 + \cdots\cdots + C_i\lambda_i\right)$$

Here λ_i are numerical constants characteristic of ions and x is a proportionality factor which accounts for the geometry of the cell. Thus, with proper calibration conductivity measurement can be related to give concentration of any electrolyte.

The conductance of a solution can be measured by determining its resistance using a Wheatstone bridge arrangement or a direct reading conductivity bridge. The reaction mixture is kept in a conductivity cell whose cell constant is computed earlier. Conductivity of the reaction mixture, kept in a thermostat, is measured at different time intervals, while a certain reaction is being continued in the cell. Suppose hydrolysis of alkyl chlorosulphate is being studied.

$$ROSO_2Cl + H_2O \rightarrow ROH + SO_3^{-2} + Cl^- + H^+$$

A reaction solution containing (0.01*M*) chlorosulphate in aqueous dioxane is taken in the cell and conductivity measurement as a function of time is made.

5.9.2 Potentiometric Measurements

Potentiometric measurements require that a particular reaction is carried out in presence of an indicator electrode, which is selective with respect to a specific chemical species present either as a reactant or product, *e.g.*, in the bromination reaction of *N, N*-dimethyl aniline (DMA)

$$C_6H_5N\,(CH_3)_2 + Br_2 \rightarrow C_6H_4N\,(CH_3)_2\,Br + HBr$$

The solution containing DMA, bromine and a large concentration of bromide ion is taken in a beaker and a platinum foil attached to a platinum wire fused in glass-tube works as an indicator electrode, while saturated calomel electrode is joined to make it an electrochemical cell. The E.M.F. of this cell is given as

$$E = E^0_{\text{cell}} - \left(\frac{2.303\,RT}{2\,F}\right)\log_{10}\left[\frac{C_{Br_2}}{\left(C_{Br^-}\right)^2}\right]$$

As the bromide ion is taken in excess, its concentration is taken as constant; we then write

$$E = E^0_{\text{cell}} + \left(\frac{2.303\,RT}{2\,F}\right)\log_{10}\left(C_{Br^-}\right)^2 - \left(\frac{2.303\,RT}{2\,F}\right)\log_{10} C_{Br_2}$$

or

$$E = E'_{\text{cell}} - \left(\frac{2.303\,RT}{2\,F}\right)\log_{10} C_{Br_2}$$

On differentiating this equation with respect to time (t),

$$\left(\frac{dE}{dt}\right) = -\left(\frac{2.303\,RT}{2\,F}\right)\left[\left(\frac{d\log_{10} C_{Br_2}}{dt}\right)\right] \tag{5.14}$$

$$= -\left(\frac{2.303\,RT}{2F}\right)\left[\left(\frac{d\,C_{Br_2}}{dt}\right)\left(\frac{1}{C_{Br_2}}\right)\right]$$

Now the rate of reaction can be given as

$$-\left(\frac{d\,C_{Br_2}}{dt}\right) = k\,C_{DMA}\,C_{Br_2}$$

This can be rearranged as

$$-\left(\frac{1}{C_{Br_2}}\right)\left(\frac{d\,C_{Br_2}}{dt}\right) = k\,C_{DMA}$$

Hence Eq. 5.14 may be presented as

$$\left(\frac{dE}{dt}\right) = \left(\frac{RT}{2F}\right) k\, C_{DMA}$$

or $$k\, C_{DMA} = \left(\frac{2F}{RT}\right)\left(\frac{dE}{dt}\right)$$

or $$k' = \left(\frac{2F}{RT}\right)\left(\frac{dE}{dt}\right)$$

Thus k', can be calculated either by carrying out potentiometric estimation of cell-EMF at various time intervals or getting a plot of cell-EMF vs time. The slope of this plot gives (dE/dt).

5.9.3 Optical Methods : Optical Rotation Method

Sucrose (table sugar) undergoes hydrolysis to produce fructose and glucose.

$$\underset{\text{sucrose}}{C_{12}H_{22}O_{11}} + H_2O \overset{H^{\ominus}}{\rightleftharpoons} \underset{\text{glucose}}{C_6H_{12}O_6} + \underset{\text{fructose}}{C_6H_{12}O_6}$$

A mixture of fructose and glucose is called invert sugar because sugar is dextro-rotatory (+), whereas the mixture of glucose and fructose resulting from inversion is laevorotatory (–). Thus a decrease in the concentration of sucrose will be accompanied by proportionate decrease in optical rotation.

Polarimetry is the measurement of the rotation of the plane of plane polarized of light. Optically active substances, *e.g.*, chiral molecules exhibit circular birefringence, which causes rotation of the polarization of the plane polarized light as it passes through the sample. A simple polarimeter instrument used to measure optical rotation consists of a long tube with flat glass ends, into which sample, *e.g.*, sucrose solution, is placed. At each end of the tube is attached a polarizer and analyser. Plane polarized light is allowed to enter through the tube, and the analyser, attached to an eye piece at the other end, is rotated until all light is shut off. The angle of rotation is then read from a scale and specific rotation of the sample/reaction mixture taken in the sample tube can be obtained with respect to time.

Problem 10. Using the data as below, show that the hydrolysis of sucrose is of first order and find k.

Time (min)	0	7.20	18.0	27.0	∞
Optical rotation (a)°	+ 24.08	+ 21.40	+ 17.73	+ 15.01	– 10.73

Solution : The total change in optical rotation from $t = 0$ to $t = \infty$, is $(\alpha_0 - \alpha_\infty)$. This is proportional to the total concentration change or initial concentration of sucrose. The change in optical rotation at time (t), *i.e.*, $(\alpha_0 - \alpha_t)$, is proportional to the change in concentration of sucrose at t. Thus, the concentration of sucrose remaining at t

$$= (\alpha_0 - \alpha_\infty) - (\alpha_0 - \alpha_t) = (\alpha_t - \alpha_\infty)$$

Since it is a first order reaction

$$\log_{10}\left[\frac{(C_{\text{sucrose}})_t}{(C_{\text{sucrose}})_0}\right] = \log_{10}\left[\frac{(\alpha_t - \alpha_\infty)}{(\alpha_0 - \alpha_\infty)}\right] = -kt$$

A plot of $\log_{10}\left[\frac{(\alpha_t - \alpha_\infty)}{(\alpha_0 - \alpha_\infty)}\right]$ vs time (t) will be linear and will have a slope of $(-k)$.

t	α	$\log_{10}\left[\frac{(\alpha_t - \alpha_\infty)}{(\alpha_0 - \alpha_\infty)}\right]$
0.00	+ 24.08	0.0000
7.20	+ 21.40	– 0.08011
18.0	+ 17.73	– 0.20141
27.0	+ 15.01	– 0.30186
∞	– 10.73	

The plot is shown below :

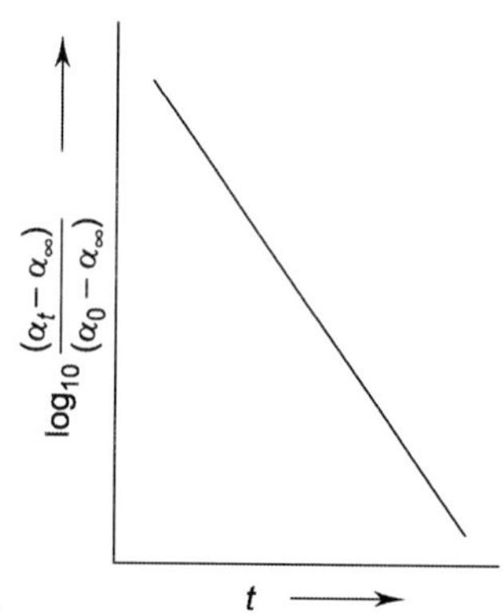

Fig. 5.4 Linear plot.

The plot is linear. Hence reaction is first order and $k = 1.11 \times 10^{-2}\ s^{-1}$.

5.9.4 Spectrophotometric Method

Spectrophotometer consists of two instruments, a spectrometer for producing light of any selected colour (wavelength) and a photometer for measuring the intensity of light. The instruments are arranged such that the liquid in a cuvette can be placed between the spectrometer beam and the photometer. The amount of light passing through the tube is measured by the photometer. The photometer delivers a voltage signal to a display device, normally a galvanometer. The signal changes as the amount of light absorbed by the liquid changes.

The development of the colour can be linked to the concentration of a substance in a solution and then the concentration can be measured by determining the extent of absorption of light of a particular wavelength.

Haemoglobin appears red because it absorbs blue and green light rays more than red. Degree of absorbance of blue or green light is proportional to the concentration of haemoglobin.

When monochromatic light of a specific wavelength passes through a solution, the relationship (Beer's Law) between the solute concentration and the intensity of transmitted light is given as

$$I = I_0 \times 10^{-kC\,l}$$

where, I and I_0 are the intensities of transmitted light using pure solvent and solution of compound in it, respectively; C is the concentration of coloured compound; l is the distance light passes through the solution; and k is a constant. In case of a spectrophotometer, the length l of light path, is taken as constant; hence the above equation may be given as

$$\frac{I}{I_0} = 10^{-k'\,C} = T$$

Here k' is a new constant and T is the transmittance of the solution. When

$$-\log_{10} T = \log_{10}\left(\frac{1}{T}\right) = kC = \text{Optical Density (OD)}$$

(OD) is thus directly related to the concentration of coloured substance.

Kinetics of base catalysed hydrolysis of *p*-nitro phenyl acetate in buffered aqueous solution has been studied spectrophotometrically. The reaction is represented as

$$p\text{-}NO_2\text{–}C_6H_4\text{–O–CO–}CH_3 + H_2O \xrightarrow{OH^-} p\text{-}NO_2\text{–}C_6H_4\text{–}O^- + O^{\ominus}OC\text{–}CH_3 + 2H^+$$

p-nitrophenylete anion (PNPA)

The hydrolysis is catalysed by OH^-, which is controlled by dilute sodium hydroxide in a buffer of pH 6.86–7.00. The rate of reaction is given as

$$-\frac{(d\,C_{ester})}{dt} = k\,C_{ester}\,C_{OH^-} = \frac{(d\,C_{PNPA})}{dt}$$

Condition of constant OH^- ion modifies this reaction to pseudo first order and on completion of reaction $(C_{PNPA}) = (C_{ester})_0$;

Hence $$\frac{(d\,C_{PNPA})}{dt} = k'\,[(C_{ester})_0 - (C_{PNPA})]$$

The reaction is followed spectrophotometrically by observing the increase in light absorption at 400 nm, caused by the formation of *p*-nitro phenylate anion (PNPA). Let A be the absorbance of reaction solution, so

$$A = \varepsilon(C_{PNPA})$$

or $$\frac{dA}{dt} = k'(A_\infty - A)$$

Here, ε is the molar absorptivity of phenolate ion. This differential equation on integration and applying the condition that when $t = 0, A = A_0$ gives

$$\log_e \left[\frac{(A_\infty - A)}{(A_\infty - A_0)}\right] = -kt$$

or $$\log_e (A_\infty - A) - \log_e (A_\infty - A_0) = -kt.$$

5.10 THEORIES OF CHEMICAL KINETICS

5.10.1 Effect of Temperature on Rate of Reaction

It can be usually thought than an increase in temperature would increase the rate of a reaction. Looking at a typical rate law, $r = k\, C_A \times B_y$, changing the temperature would not effect the concentrations of A and B, so any effect of temperature must be looked as an effect on k. Thus, k must increase with increasing temperature.

The effect of temperature on reaction rate constants can be of several kinds:

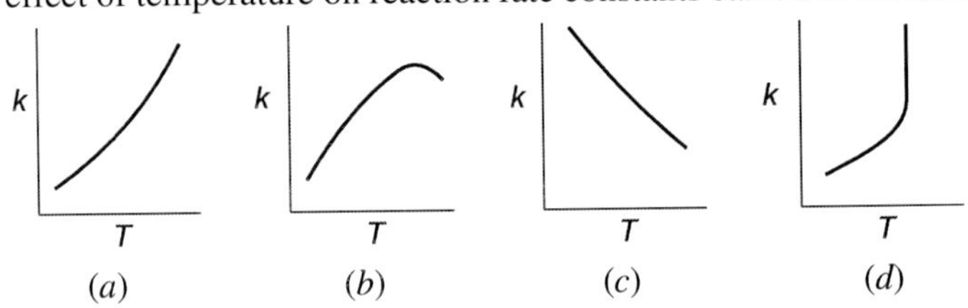

Fig. 5.5 Effect of temperature on rate constants.

Plot (a) shows rate constants increase with increasing temperatures. Plot (b) follows a similar trend, a temperature is reached where the rate constant approaches maximum and then starts to decrease with further increase in temperature. This is what happens in an enzyme-catalysed reaction. As long as the enzyme remains in its native state, the rate constant increases with increasing temperature, but eventually a temperature is reached where the enzyme gets denatured, and the rate constant starts to decrease. Plot (c) has been observed for reactions where one of the steps in the reaction is strongly exothermic and this effect influences heavily on the increment in the rate constant. An example of the reaction is $N_2 + O_2 \rightarrow 2NO$. Plot (d) illustrates the trend in rate constants for chain reactions. The rate gradually increases with temperature until a threshold is reached where the propagation reactions become significant. Any further increase in temperature causes an uncommon increase in the rate constant.

A rough rule for reactions in solution near room temperature is that k increases 2–3 times for every 10°C increase in temperature. The ratio of rate constant at 35°C and 25°C is knoun as temperature coefficient and is equal to appriximately 2 for most common reations.

$$\frac{k_{35}}{k_{25}} \simeq 2 \quad \text{(Temperature coefficient)}$$

Most commonly the rate goes up with temperature, but this is not always the case. For example, the rate constant for the reaction

$$2NO + O_2 \longrightarrow 2NO_2$$

decreases with increase in temperature.

Experimentally, it has been found that in a very large number of case the rate constants, with some level of approximation, vary with temperature according to the **Arrhenius equation:**

$$k_{(T)} = A \exp^{\left(\frac{E_a}{RT}\right)} \quad (5.15)$$

where the rate constant has been presented as $k_{(T)}$ to emphasize that it depends on temperature, R is the gas constant and has dimensions of energy temperature^{-1} mol^{-1}, so the product RT has dimensions of energy mol^{-1}. E_a has units of energy mol^{-1} (kJ mol^{-1}) and is called the **activation energy**.

Typical values for E_a are between 10 and 200 kJ mol^{-1}, for example for

$$2\ N_2\ O_5 \rightleftharpoons 2\ N_2\ O_4 + O_2;\ E_a = 160\ \text{kJ mol}^{-1}$$

A is called the frequency factor. It is often found to be independent of temperature, or at least only weakly dependent on temperature. Since the exponential term is dimensionless, A must have the same dimensions and units as k.

The Arrhenius equation predicts that, for a positive activation energy, the rate constant increases with temperature. The equation can be manipulated into a straight line form by taking natural logarithms of both sides :

$$\log_e k = \log_e A - \left(\frac{E_a}{RT}\right)$$

So a plot of $\log_e k$ against $\frac{1}{T}$ should be a straight line ; such a plot is called an **Arrhenius plot** Fig. 5.6. The slope of the plot is $-\frac{E_a}{R}$ (dimensions temperature, *e.g.*, K, and the intercept with the *y*-axis, when $\frac{1}{T}$ goes to zero, is $\log_e$ A. Thus activation energy (E_a) can be determined from the slope of this straight line plot.

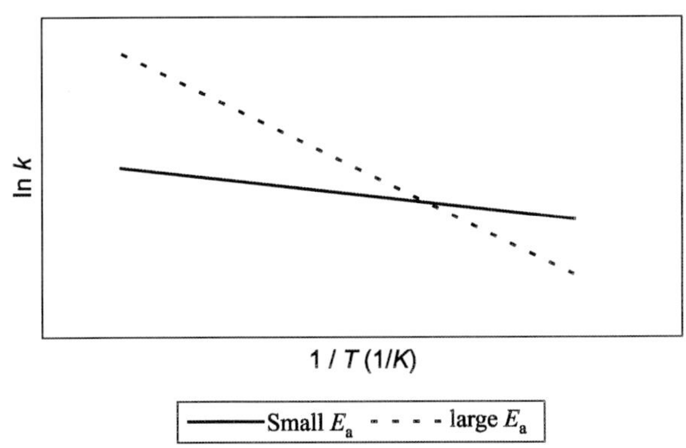

Fig. 5.6 Arrhenius plot of $\log_e k$ vs $1/T$.

Many chemical processes show Arrhenius behaviour, even if the processes themselves turn out to be rather complex.

5.10.2 Equilibrium Constant and Rate Constants

Chemical equilibrium is dynamic in nature. The forward and back reactions continue just as they do when a system is away from equilibrium, but when equilibrium has been reached these forward and backward rates are equal. As a result, there is no change in the concentration of any of the species, even though the reactions are still going on.

Suppose we have a reaction of the form

$$A + B \rightleftharpoons C + D$$

The rate of forward reaction will be $k_f\, C_A\, C_B$ and rate of backward reaction will be $k_b\, C_C\, C_D$ where k_f and k_b are respective rate constants for forward and backward reactions. At equilibrium two rates are equal

$$k_f\, C_A\, C_B = k_b\, C_C\, C_D$$

or

$$K_{eq} = \left(\frac{k_f}{k_b}\right) = \frac{(C_C\, C_D)}{(C_A\, C_B)}$$

The equilibrium constant, K_{eq} is the ratio of the rate constants of the forward (k_f) and backward (k_b) reactions.

Differentiating the logarithmic form of Arrhenius equation with respect to T, we have

$$\frac{(d \log_e k)}{dT} = \left(\frac{E_a}{RT^2}\right)$$

Frequency factor A, being independent of T, may be taken as constant. Integrating this equation between T_1 and T_2, when the corresponding rate constants are k_1 and k_2, respectively. The term E_a, activation energy, if remains unchanged during temperatures T_1 and T_2, may be taken as constant. Thus,

$$\log_e \left(\frac{k_2}{k_1}\right) = \left(\frac{E_a}{R}\right)\left[\frac{(T_2 - T_1)}{T_1\, T_2}\right]$$

or

$$\log_{10} \left(\frac{k_2}{k_1}\right) = \left(\frac{E_a}{2.303\, R}\right)\left[\left(\frac{1}{T_1} - \frac{1}{T_2}\right)\right] \tag{5.16}$$

This is the integrated form of **Arrhenius equation**. The energy of activation for a reaction can be evaluated using this equation, if we know the rate constants for the reaction at two different temperatures.

5.10.3 Activation Energy (E_a)

The concept of activation energy emphasizes that all the molecules of the reactants can not take part in chemical reaction. Only activated molecules can take part in the chemical reaction. Thus, the reactant molecules must first acquire the necessary energy to pass through an energy barrier or activated state or transition state. This amount of energy, which the reactant molecules have to absorb so that they can become activated and pass through energy barrier, is called the **activation energy.**

This necessary energy of activation is acquired by the reacting molecules during their collisions, when the molecules with higher energy interchange energy with molecules of lower energy.

Now when two activated molecules, possessing needed activation energy, come close to each other, they collide and first form an activated complex or transition state, which eventually lead to products.

Relation between activation energy and the enthalpy of a reaction can be better understood through the energy barrier concept Fig. 5.7. Basically a reaction is the transformation of the reactant system, present at a certain potential energy state, to product system present at another potential energy state, accompanying the heat change ΔH. This transformation from reactant state to product state takes place by an input of energy E_a and the reverse process by an expenditure of energy (E_b). Thus, $[E_a - E_b] = \Delta H$, the enthalpy of reaction at constant temperature. Hence it can be observed that reactant molecules have to cross the energy barrier, permitting them to reach at transition state, before they can arrive to the product state. This height of energy barrier is the energy of activation, E_a Pathway of minimum energy from reactants to product is called the reaction co-ordinate.

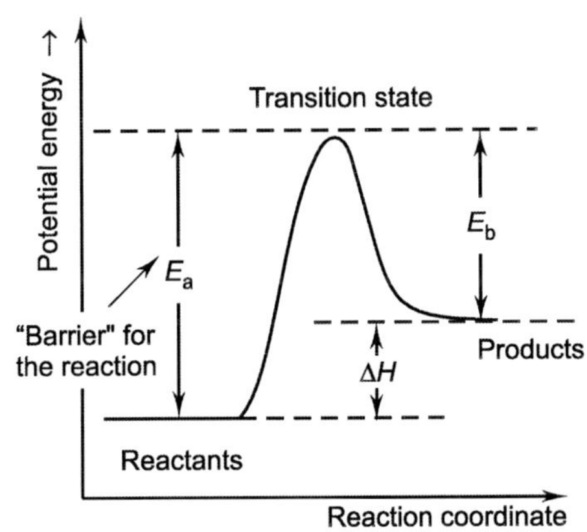

Fig. 5.7 Energy change in a chemical reaction.

5.10.4 Simple Collision Theory Based on Hard Sphere Model

Consider an elementary biomolecular reaction

$$A + B \longrightarrow \text{Products}$$

This theory suggests that for the reaction between A and B to occur, they must collide with some minimum energy, *i.e.*, equal to activation energy. Hence, the rate of reaction is proportional to the fraction of collisions of such molecules which possess kinetic energy equal to their activation energy, E_a. The exponential term in Arrhenius equation accounts for the same. Rate of reaction is also proportional to the frequency of such collisions. There is also a pre-requisite that for successful collisions, molecules must be properly oriented, a steric factor, being reflected by a pre-exponential factor in Arrhenius equation, accounts for this. For example, a reaction between a fluorine molecule and nitrogen dioxide requires that the fluorine molecule attacks from the rear; an attack from the front is much less likely to be successful.

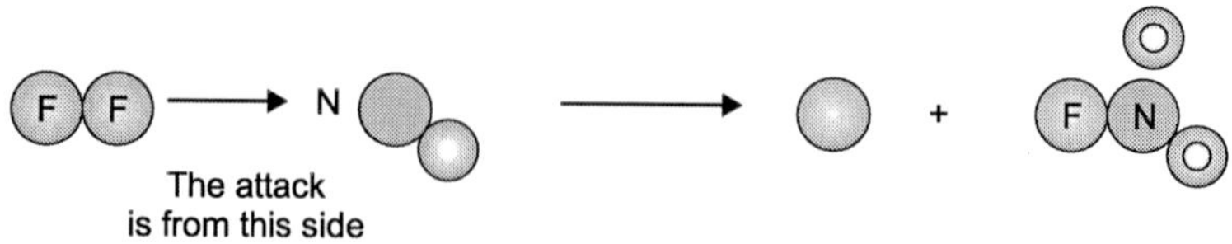

$$F_2 + NO_2 \longrightarrow F + FNO_2$$

The number of binary collisions between two unlike molecules (A and B), having concentration as C_A and C_B molecules per litre, can be given as

$$Z_{AB} = \left[\frac{8\pi RT\left(M_A + M_B\right)}{M_A M_B}\right]^{\frac{1}{2}} \Sigma_{AB}^2 C_A C_B$$

Here M_A and M_B are the molecular weights and Σ_{AB} is the mean of the molecular diameters and called effective molecular diameter. The number of effective collisions between A and $B = Z_{AB}\, q$, where q is the fraction of total molecules which possess energy equal to or greater than the critical value E_{coll}.

Fraction q is given by Maxwell's distribution law as

$$q = \exp^{-E_{coll}/RT}$$

Hence, reaction rate

$$= \left[\frac{8\pi RT\left(M_A + M_B\right)}{M_A M_B}\right]^{\frac{1}{2}} \left(\Sigma_{AB}^2 C_A C_B\right)\exp^{-E_{coll}/RT}$$

For a bimolecular reaction,

Reaction rate = $k\, C_A\, C_B$

where, k is the specific reaction rate. Hence,

$$k = \Sigma_{AB}^2 \left[\frac{8\pi RT\left(M_A + M_B\right)}{M_A M_B}\right]^{\frac{1}{2}} \exp^{-E_{coll}/RT} \ \text{l molecule}^{-1}\ \text{s}^{-1}$$

or

$$k = \left(\frac{N}{1000}\right)\Sigma_{AB}^2 \left[\frac{8\pi RT\left(M_A + M_B\right)}{M_A M_B}\right]^{\frac{1}{2}} \exp^{-E_{coll}/RT} \ \text{dm}^3\ \text{mol}^{-1}\ \text{s}^{-1}$$

Here N is the Avogadro's number.

Putting
$$A = \left(\frac{N}{1000}\right)\Sigma_{AB}^2 \left[\frac{8\pi RT\left(M_A + M_B\right)}{M_A M_B}\right]^{\frac{1}{2}}$$

$$K = A\, p\, \exp^{-E_a/RT} \tag{5.17}$$

Here E_{coll} is identified with the Arrhenius Activation energy, E_a, and a steric factor, p, has been introduced to account for orientational requirements. Thus, we get a generalized rate constant based on collision theory.

Weaknesses of collision theory

The weaknesses of collision theory are as follows :

(*a*) The assumption that reactant molecules behave as rigid spheres is unacceptable rather molecular structure should have been taken into consideration.

(*b*) It has been applied only to simple homogeneous reactions.

(*c*) Internal motion of the molecule has not been accounted.

5.10.5 Transition State Theory (Equilibrium Hypothesis)

Transition state theory or activated complex theory was developed by H. Eyring and M. Polanyi in the 1930s. It is applicable to gas and solution phase reactions. It has been assumed in this theory that reactants form a transition state or activated complex which is in equilibrium with them and it then decomposes to yield the product.

$$\underset{\text{Reactants}}{M + N} \underset{}{\overset{\boldsymbol{K}^\ddagger}{\rightleftharpoons}} \underset{\substack{\text{Activated}\\\text{Complex}}}{MN^\ddagger} \xrightarrow{K^\ddagger} \text{Products}$$

The transition state $MN^\ddagger$ has a life-time of only a few molecular vibrations and can be observed by **fento-second (10^{-15} s) or pico-second (10^{-12} s) spectroscopy**. $\boldsymbol{K}^\ddagger$ is the equilibrium constant between the reactants and transition state, whereas $K^\ddagger$ is the universal rate constant. The equilibrium constant $\boldsymbol{K}^\ddagger$ can be represented as

$$\boldsymbol{K}^\ddagger = \frac{\left[C_{MN^\ddagger}\right]}{\left[C_M\, C_N\right]}$$

and the rate of formation of product with the decomposition of activated complex as

$$\frac{dP}{dt} = K^\ddagger\, [C_{MN}{}^\ddagger]$$

These are then expressed as

$$\frac{dP}{dt} = K^\ddagger\, \boldsymbol{K}^\ddagger\, [C_M\, C_N]$$

where the rate constant $k = K^\ddagger\, \boldsymbol{K}^\ddagger$. The activated complex would decompose only if enough vibrational energy is supplied to the system so as to cause the atoms to vibrate with certain critical frequency leading to the bond rupture. This frequency (ν) is given by

$$\nu = \frac{E_{vib}}{h}$$

Average energy of vibration is given by

$$E_{vib} = k\,T = \left(\frac{R}{N_A}\right) T = k_B\,T$$

Here N_A is the Avogadro's number and k_B is Boltzmann constant. The vibrational frequency (ν) is the rate at which the activated complex molecules move across the energy barrier. Hence, it is identified with the rate constant = $K^{\ddagger}$. Thus,

$$\frac{dP}{dt} = K^{\ddagger}\,\boldsymbol{K}^{\ddagger}\,[C_M\,C_N]$$

$$= \left(\frac{R}{N_A}\right)\left(\frac{T}{h}\right)\,\boldsymbol{K}^{\ddagger}\,[C_M\,C_N]$$

Since the rate of reaction $= k\,[C_M\,C_N]$

Hence, the specific reaction rate, k is expressed as

$$k = \left(\frac{R}{N_A}\right)\left(\frac{T}{h}\right)\boldsymbol{K}^{\ddagger} = \left(\frac{k_B T}{h}\right)\boldsymbol{K}^{\ddagger}$$

This is the general expression for the rate constant of any elementary process in the transition state theory. Thus, $\boldsymbol{K}^{\ddagger}$ can be determined experimentally from the rate constant.

5.10.6 Thermodynamic Treatment of Transition State Theory

We know from thermodynamics that for any equilibrium process, the equilibrium constant **($\boldsymbol{K}^{\ddagger}$) is related to the Free Energy Change ($\Delta G^{\ddagger}$) as**

$$(\Delta G^{\ddagger}) = -RT\log_e \boldsymbol{K}^{\ddagger}$$

or

$$\boldsymbol{K}^{\ddagger} = \exp^{-\Delta G^{\ddagger}/RT}$$

But $\Delta G^{\ddagger}$ is the difference between the free energy of the activated complex and that of the reactants in their standard states. Also,

$$\Delta G^{\ddagger} = \Delta H^{\ddagger} - T\,\Delta S^{\ddagger}$$

Hence, specific reaction rate can be expressed as

$$k = \left(\frac{R}{N_A}\right)\left(\frac{T}{h}\right)\boldsymbol{K}^{\ddagger} = \left(\frac{R}{N_A}\right)\left(\frac{T}{h}\right)\exp^{-\Delta G^{\ddagger}/RT}$$

or

$$k = \left(\frac{R}{N_A}\right)\left(\frac{T}{h}\right)\exp\left(\frac{.\Delta S^{\ddagger}}{R}\right)\exp^{-\Delta G^{\ddagger}/RT}$$

where $\Delta S^{\ddagger}$ and $\Delta H^{\ddagger}$ are the entropy and the enthalpy of activation.

Comparing this with Arrhenius equation

$$k\,(T) = \mathrm{A}\,\exp^{E_a/RT}$$

We can write

$$A \approx \left(\frac{R}{N_A}\right)\left(\frac{T}{h}\right)\exp^{\Delta S^{\ddagger}/R}$$

and $$E_a \approx -\Delta H^{\ddagger}$$

Thus, when activation energy of a reaction is experimentally determined it can be used to compute enthalpy and entropy changes for the reaction.

5.11 COMPLEX REACTIONS

There are many reactions which involve two or more side reactions or undesirable reactions in addition to the reaction being studied. Such undesirable reactions complicate the study of the kinetics of the main reaction. Following are some of the factors which cause such problems :

(*i*) Parallel or side reactions

(*ii*) Reversible reactions

(*iii*) Chain or Consecutive reactions

(*i*) **Parallel or Side Reactions:** Such reactions are those in which reactants combine in more than one way to yield different products. For example,

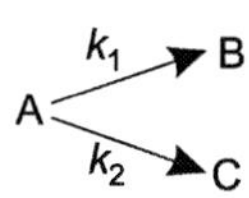

(main reaction)

(side reaction)

The rate of disappearance of A will be the sum of the rates of both the reactions

$$-\frac{d[A]}{dt} = k_1\,[\mathrm{A}] + k_2\,[A]$$

B and C will be produced in the ratio of rate constants

$$\frac{d[B]}{d[C]} = \frac{k_1}{k_2}$$

Most organic reactions have a side reaction. For example,

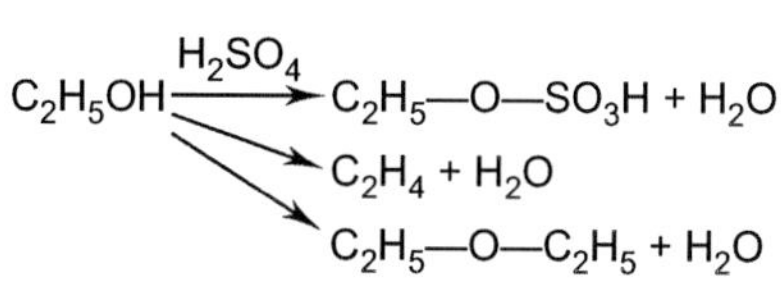

(*ii*) **Reversible Reactions:** A reaction in which products recombine to form original reactants is a reversible reaction.

For a reaction $A \underset{k_{-1}}{\overset{k_1}{\rightleftharpoons}} B$, the rate is given by

$$\frac{dx}{dt} = k_1 (a - x) - k_{-1} (x)$$

if a is the initial concentration of A and x is the concentration of B after time t, at equilibrium the rates of forward and backward reaction become equal

$$k_1 (a - x) = k_{-1} (x)$$

and
$$\frac{k_1}{k_{-1}} = k_c \text{ (equilibrium constant)}$$

(*iii*) **Chain reaction:** Such reactions take place in a series of successive steps involving reactive species (which may be atoms or free radicals) before the formation of final product.

For example, the reaction between H_2 and Br_2 to form HBr involves the following steps :

Step 1. $Br_2 \xrightarrow{k_1} 2Br$] chain initiation

Step 2. $Br + H_2 \xrightarrow{k_2} HBr + H$ } chain propagation

Step 3. $H + Br_2 \xrightarrow{k_3} HBr + Br$

Step 4. $2Br \xrightarrow{k_4} Br_2$ } chain termination

Step 5. $H + HBr \xrightarrow{k_5} H_2 + Br$

Since HBr is involved in steps 2, 3 and 5,

$$\frac{d[HBr]}{dt} = k_2 [Br] [H_2] + k_3 [H] [Br_2] - k_5 [H] [HBr]$$

Such chain reactions may lead to explosion.

5.12 CATALYSIS

A catalyst is a substance which alters the reaction rate without being spent during the chemical transformation.

J.J. Berzélius identified the phenomenon of catalysis in 1835 and W. Ostwald recognized its kinetic origin in 1911.

Catalysis May be of two Types: (*i*) positive catalysis – where the catalyst increases the rate of reaction ; and (*ii*) negative catalysis – when it decreases the rate of reaction.

The catalyst does not appear in the chemical equation, although it intervenes in the reaction mechanism and its action results in a change in activation energy. In a positive catalysis reaction, the catalysed path requires activation energy E_a (cat) much lower than activation energy, E_a, of reaction taking place in absence of catalyst.

A catalyst may open new reaction paths leading to other reaction products.

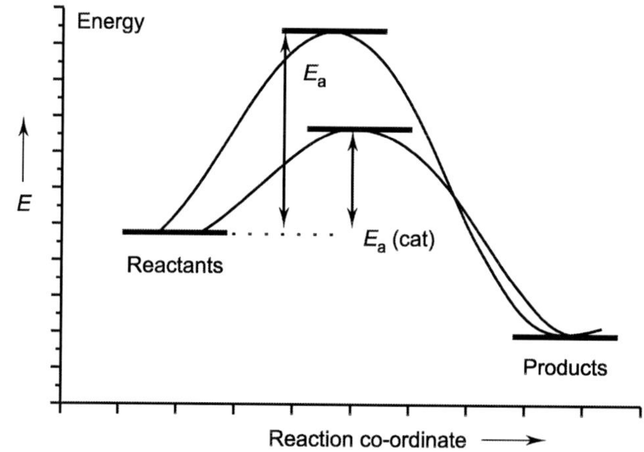

Fig. 5.8 Activation energy of a catalysed reaction.

5.12.1 Characteristics of Catalysed Reactions

Important characteristics of catalysed reactions are :

(*i*) A catalyst remains chemically unchanged during the course of a reaction; it may undergo a change in its physical form, *e.g.*, a lot of roughness being developed in platinum gauge utilized as a catalyst in ammonia oxidation in the manufacture of nitric acid.

(*ii*) Free energy change (ΔG) for a reaction at a definite temperature remains the same, whether it takes place in presence of a catalyst or in its absence. As ΔG is related to equilibrium constant of a catalysed reaction, *i.e.*, $-(\Delta G) = R\text{T} \log_e K$, equilibrium constant remains unchanged. A catalyst, thus offers a rapid attainment of equilibrium position; however, the relative proportion of reactants and products remains unaffected.

(*iii*) For a reaction to occur there must be a decrease of free energy and since catalyst does not contribute in any way to the total free energy of the system, it cannot initiate a reaction.

(*iv*) Most catalysts are specific in their action. When ethyl alcohol vapours are passed at 350°C over nickel surface it gets dehydrogenated :

$$C_2H_5OH \xrightarrow{Ni} CH_3CHO + H_2$$

On the other hand passing them over heated alumina, ethanol gets dehydrated :

$$C_2H_5OH \xrightarrow{Al_2O_3} C_2H_4 + H_2O$$

(*v*) A very small amount of catalyst is needed to catalyse a reaction. A concentration of 1 g ion of Cu (II) in 10^6 l of sodium sulphite solution is sufficient to catalyse its oxidation.

(*vi*) Functioning of catalyst is optimum at a definite temperature.

(*vii*) Traces of impurities inhibit or retard the rate of a catalysed reaction. These substances, which either inhibits or slows down the catalyst functioning, are called **catalytic poisons.** Traces of mercury reduce the catalytic activity of copper in the formation of ethane from ethylene and hydrogen.

(*viii*) Presence of small amounts of certain substances, which themselves are not catalyst, may enhance the functioning of a catalyst. These are called catalytic **promoters**. Haber's process for the manufacture of ammonia utilizes finely divided iron as catalyst and molybdenum as promoter. A promoter alone cannot catalyse the reaction.

5.12.2. Classification of Catalysis

There are three types of catalysis :

(*a*) Homogeneous catalysis
(*b*) Heterogeneous catalysis and
(*c*) Enzyme catalysis.

(*a*) HOMOGENEOUS CATALYSIS

The catalyst and the reagents are in homogeneous phases. The catalyst molecule (*C*) enters into formation of a complex (*X*) with one of the reactant (substrate), *S*, in the first step of this type of catalysed reaction. This complex may either react with another molecule, *N*, or itself undergo subsequent decomposition leading to the formation of product (*P*) along with some elimination

$$C + S \underset{k_{-1}}{\overset{k_1}{\rightleftharpoons}} X + M$$

$$X + N \xrightarrow{k_2} P + Z$$

When C_{C_0} and C_{S_0} are the initial concentrations of catalyst and substrate respectively, we can write the concentration of catalyst, C_c, and substrate, C_s as

$$C_c = \left(C_{C_0} - C_X\right) \text{ and } C_s = \left(C_{S_0} - C_X\right)$$

If reaction is occurring with an appreciable rate and catalyst is taken in very small concentration, steady-state treatment can be applied for concentration of complex, C_X, *i.e.*,

$$\frac{dC_X}{dt} = k_1\left[\left(C_{C_0} - C_X\right)\left(C_{S_0} - C_X\right)\right] - k_{-1}\, C_X\, C_M - k_2\, C_X, C_N = 0$$

Since C_X is quite small, term C_X^2 can be neglected, and therefore

$$C_X = \frac{\left(k_1\, C_{C_0}\, C_{S_0}\right)}{\left[k_1\left(C_{C_0} + C_{S_0}\right) + k_{-1}\, C_M + k_2\, C_N\right]}$$

Hence the rate of reaction could be written as

$$-\frac{dC_s}{dt} = k_2\, C_X\, C_N$$

$$= \frac{\left(k_1\, k_2\, C_{C_0}\, C_{S_0}\, C_N\right)}{\left[k_1\left(C_{C_0} + C_{S_0}\right) + k_{-1}\, C_M + k_2\, C_N\right]}$$

Thus, at low substrate and catalyst concentrations, the rate of reaction is proportional to the first power of both the catalyst and substrate concentrations. On the other hand when substrate is present at high concentration the rate become independent of its concentration. Variation of rate with substrate concentration, C_{S_O}, for a homogeneous catalysis can be represented as in Fig. 5.9.

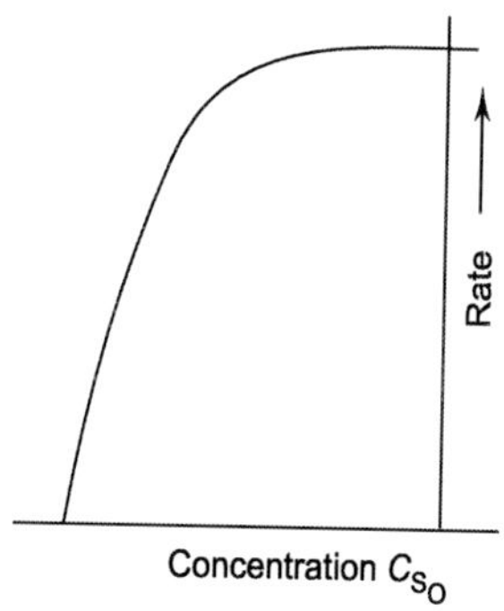

Fig. 5.9 Rate of homogeneous catalysis vs substrate concentration.

As the catalyst effectively increases the rate of formation of intermediate complex, the potential energy and thus energy of activation of catalysed reaction, E_a (cat), lies much below the value of energy of activation of reaction (E_a) taking place without catalyst.

■ **Example 1.** Chlorine radicals ($Cl^{\cdot}$), usually formed by the decomposition of chlorofluorocarbons through UV radiations of sunlight, catalyses the decomposition of stratospheric ozone by forming with it an intermediate complex radical ($ClO^{\cdot}$) in the primary first stage. In the next stage these intermediate radicals regenerate chlorine radicals by reacting with the oxygen atom

$$Cl^{\cdot} + O_3 \longrightarrow ClO^{\cdot} + O_2$$
$$ClO^{\cdot} + O^{\times} \longrightarrow Cl^{\cdot} + O_{23}$$

■ **Example 2.** Oxidation of sulphur dioxide is catalysed by nitric oxide, which forms an intermediate compound, nitrogen dioxide, in presence of oxygen. This intermediate compound then reacts with sulphur dioxide forming the product sulphur trioxide, and regenerating catalyst molecule.

$$NO + \frac{1}{2} O_2 \longrightarrow NO_2$$

$$NO_2 + SO_2 \longrightarrow NO + SO_3$$

Acid–Base Catalysis

Bronsted acids and bases have tendencies to lose and accept a proton, respectively. These include catalysts such as H_3O^+ and OH^- ions.

General mechanism for an acid–base catalysis is

$$\text{SUBSTRATE }(S) + \text{ACID (HA)} \underset{k_{-1}}{\overset{k_1}{\rightleftharpoons}} SH^+ + A^-$$

$$SH^+ + H_2O \xrightarrow{k_2} \text{Product }(P) + H_3\,O^+$$

Acid *HA* transfers a proton to the substrate forming intermediate complex SH^+ and conjugate base anion A^-. This protonated substrate then transfers the proton to the solvent, water, forming the product. The steady-state concentration of SH^+ could be obtained as

$$k_1\, C_s\, C_{HA} = k_{-1}\; C_{SH^+}\, C_{A^-} + k_2\, C_{SH^+}$$

or $$k_1\, C_s\,(C_{HA}) = k_{-1}\; C_{SH^+}\, C_{A^-} + k_2\, C_{SH^+}$$

or $$C_{SH^+} = \frac{k_1\, C_s\,(C_{HA})}{\left[k_{-1} C_A + k_2\right]}$$

The rate equation for the reaction is

$$\text{Rate} = k_2\, C_{SH^+}\, C_{H_2O}$$

$$= \frac{k_2\, k_1\, C_s\, C_{HA}}{\left[k_{-1} C_A + k_2\right]}$$

When $$k_2 >> k_{-1}\, C_A$$

$$\text{Rate} = k_1\, C_s\, C_{HA}$$

Hence, reaction rate is proportional to the first power of concentration of substrate and also acid (catalyst). On the other hand if $(k_{-1}\, C_A) >> k_2$

$$\text{Rate} = \frac{k_2\, k_1\, C_S\, C_{HA}}{\left(k_{-1}\, C_A\right)}$$

Considering the dissociation of acid *HA*

$HA \rightleftharpoons H^+ + A^-$; hence dissociation constant of the acid,

$$K_a = \frac{\left(C_{H^+}\, C_{A^-}\right)}{C_{HA}}$$

or
$$\frac{C_{\mathrm{H^+}}}{K_a} = \frac{C_{\mathrm{HA}}}{C_{\mathrm{A^-}}}$$

Hence,
$$\text{Rate} = \left(\frac{k_2 k_1}{k_{-1}}\right)\left(\frac{C_{\mathrm{H^+}}}{K_a}\right) C_s$$

$$\text{Rate} = k' \; C_{\mathrm{H^+}} \; C_s$$

Hence, in this case reaction rate is proportional to $C_{\mathrm{H^+}}$, even when catalyst acid HA^+ and anion A^- are in the reaction.

(B) HETEROGENEOUS CATALYSIS

Here, the catalyst is often a solid whereas the reactive phase is either a solution or a gas. This type of catalysis is very much important in industrial processes.

The reagents are fixed on the surface of catalyst usually through coordination complex formation, *i.e.*, non-covalent interactions.

Example 1. Transformation of a double C=C bond into single bond. The reaction is catalysed by a surface of palladium adsorbed on carbon:

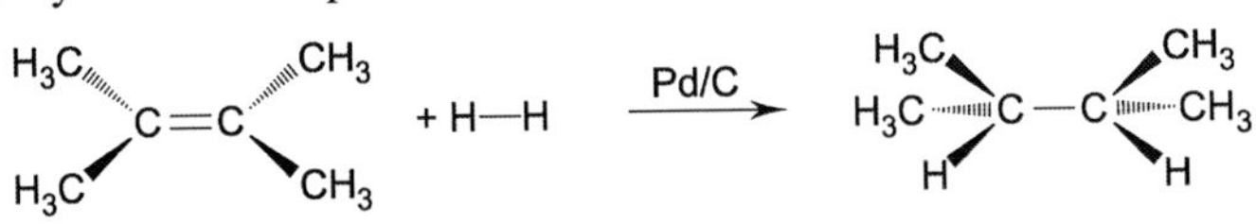

Tetramethylethene Hydrogen Tetramethylethane

The reaction occurs in several steps :

Step 1: This involves the adsorption of hydrogen and also tetramethylethene on the solid Pd–carbon fibre.

Step 2: This gives sufficient time for the reactants to combine, *i.e.*, addition of the hydrogen atoms at double bond.

Step 3: Finally desorption of the end product takes place from the catalyst surface.

Reduction of an alkyne of the type R—C≡C—R with hydrogen in presence of Pd deactivated with $BaSO_4$ (Lindlar's catalyst) to yield a *cis*-alkene is a proof of adsorption of alkyne on the surface followed by reaction with hydrogen only on one side of alkyne.

Step 1: Adsorption of alkyne

R—C≡C—R
...
...
...
———————
$Pd/BaSO_4$

Step 2: Reaction with H_2

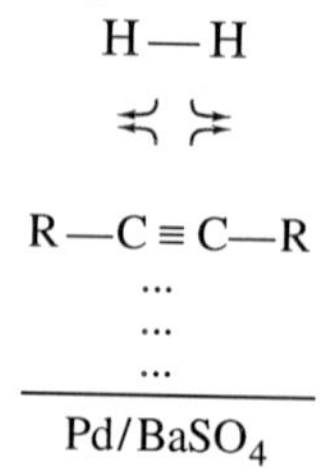

Step 3: Desorption of products

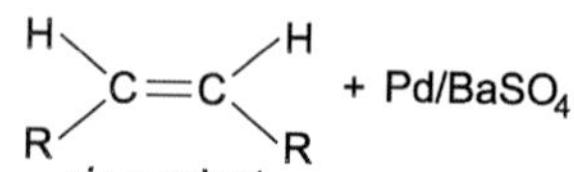

cis-product

Thus in these surface catalytic reactions the reacting substances, say P and Q, are first adsorbed on solid catalyst surface forming a non-covalent interaction. These are physical processes (Langmuir monolayer adsorption) and reaction rate on unit surface area is governed by the fraction of the catalyst surface already covered by the reacting molecules (θ). However, the adsorbed chemical substances undergo chemical reaction at the catalyst surface, which is a slow rate determining step.

$$P + S \underset{k_{-1}}{\overset{k_1}{\rightleftharpoons}} PS$$

$$Q + S \underset{k_{-2}}{\overset{k_2}{\rightleftharpoons}} QS$$

$$PS + QS \xrightarrow{k_3} \text{Products} + S$$

Thus, rate of reaction is given by

$$\text{Rate} = k_3\, \theta_P\, \theta_Q$$

Here θ_P and θ_Q are the fractions of the surface of catalyst covered by reacting molecules P and Q respectively.

Under steady-state conditions, concentrations of surface co-ordination complexes C_{PS} and C_{QS} can be evaluated as

$$\frac{(dC_{PS})}{dt} = k_1\, C_P\, C_S - k_{-1}\, C_{PS} - k_3\, C_{PS}\, C_{QS} = 0$$

$$\frac{(dC_{QS})}{dt} = k_2\, C_Q\, C_S - k_{-2}\, C_{QS} - k_3\, C_{PS}\, C_{QS} = 0.$$

Here C_S is the concentration of the vacant sites on the surface. The fraction of the vacant sites, θ_v can be given as

$$\theta_v = 1 - (\theta_P + \theta_Q)$$

Now suppose total concentration of the sites on the surface is $\hat{S}$, then

$$C_s = \hat{S}\ \theta_v,\ C_{PS} = \hat{S}\ \theta_P \text{ and } C_{QS} = \hat{S}\ \theta_Q$$

Using these expressions, we can write equations as

$$k_1 C_P \hat{S} \theta_v - k_{-1} \hat{S} \theta_P - k_3 \hat{S}^2 \theta_P \theta_Q = 0$$

$$k_2 C_Q \hat{S} \theta_v - k_{-2} \hat{S} \theta_Q - k_3 \hat{S}^2 \theta_P \theta_Q = 0$$

These equations can be solved to give us values of q_P and q_Q.

When constant k_3 is very small, then we get the final solutions in the form

$$\text{Rate} = \frac{(k_3 K_1 K_2 C_P C_Q)}{(1 + K_1 C_P + K_2 C_Q)^2}$$

where, $$K_1 = \frac{k_1}{k_{-1}} \text{ and } K_2 = \frac{k_2}{k_{-2}}$$

Now we have the following situations

Type I : Both the substrate P and Q are adsorbed to very few sites

$$K_1 C_P >> 1 \text{ and } K_2 C_Q >> 1$$

Here $K = k_3 K_1 K_2$ and if concentrations are represented in terms of partial pressure,

$$\text{Rate} = K p_P p_Q$$

Thus, reaction is of first order with respect to each of the reactants and overall order is 2.

Type II : When one of the substrate, say P, is more strongly absorbed than other (Q).

$$K_1 p_P >> K_2 p_Q$$

Hence $$\text{Rate} = \frac{K C_P C_Q}{(1 + K_1 C_P)^2}$$

The reaction is of first order with respect to Q, the substrate which is weakly absorbed with respect to the other.

Type III : When one of the substrates is very strongly absorbed. The suppose P is strongly absorbed component,

Then $$K_1 C_P >> K_2 C_Q \text{ and } K_1 C_P >> 1$$

Thus $$\text{Rate} = \frac{(k_3 K_1 K_2 C_P C_Q)}{(K_1 C_P)^2}$$

$$= \frac{(k_3 K_2 C_Q)}{(K_1 C_P)}$$

$$\text{Rate} = K' \frac{C_Q}{C_P}.$$

Hence, in this type, the rate is inversely proportional to the concentration/partial pressure of strongly absorbed component.

(*C*) ENZYME CATALYSIS

Many chemical reactions which occur inside a living cell often proceed with great efficiency and speed. The food, we eat, like rice and bread gets digested through oxidative metabolism, in an hour in the cell. The same process, if carried out in laboratory with the materials discussed would take months to achieve same efficiency and speed.

The body cells carry out similar processes with optimal speed and efficiency by using complex protein molecules, called **enzymes**, as catalyst. Thus, enzymatic catalysis is (*i*) **highly specific** *e.g.*, lactose dehydrogenase catalyses the dehydrogenation of lactic acid to form pyruvic acid only; (*ii*) **very efficient** number of substrate molecules converted / minute may be as high as 10^6.

The enzyme has a reactive site which binds the substrate by non-covalent interactions, such as hydrogen bonds, dipole–dipole interactions, dispersion forces–π stacking interactions, hydrophobic effect, *etc.*

Enzymatic process, like heterogeneous catalysis, proceeds in steps such as substrate fixation on reactive site, chemical reaction on enzyme catalyst and desorption of the product (*s*) with the regeneration of enzyme catalyst as shown in Fig. 5.10.

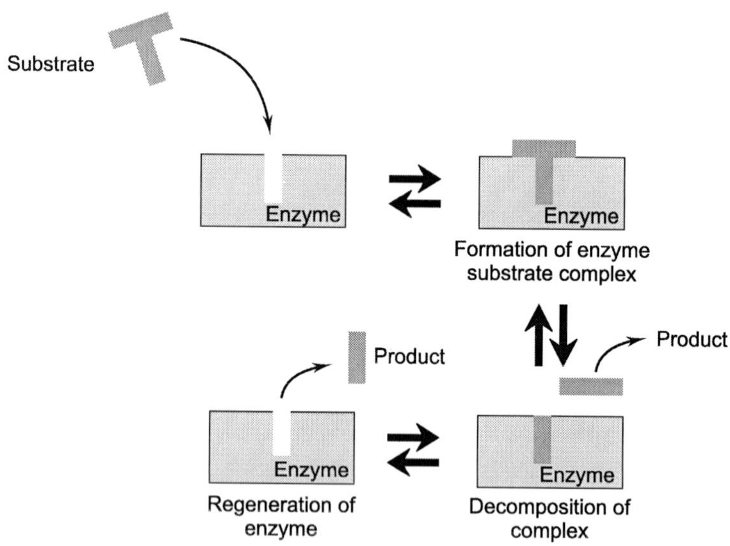

Regeneration of enzyme Decomposition of complex

Fig. 5.10 Mechanism of enzyme action.

Simple kinetic scheme, as proposed by Michaelis and Menton, using the steady state approximation is

$$E + S \underset{k_{-1}}{\overset{k_1}{\rightleftharpoons}} ES \xrightarrow{k_2} P + E$$

where, E is the enzyme, S the substrate, ES the enzyme–substrate complex and P is the product.

The three elementary steps are :

1. Formation of enzyme (E) and substrate (S) complex
2. Reversible decomposition of enzyme–substrate complex
3. Decomposition of complex into product and regeneration of enzyme.

Here it is assumed that ES is in the steady state and, as it is formed in step (1) but lost in steps (2) and (3); it follows that

$$\frac{(d\,C_{ES})}{dt} = k_1\,C_E\,C_S - k_{-1}\,C_{ES} - k_2\,C_{ES} = 0$$

or
$$C_{ES} = \frac{(k_1\,C_E\,C_S)}{(k_{-1} + k_2)}$$

Rate of product formation is given by the rate of loss of complex, *i.e.*,

$$-\frac{(d\,C_{ES})}{dt} = k_2\,C_{ES} = k_2\,\frac{(k_1\,C_E\,C_S)}{(k_{-1} + k_2)}$$

We know the initial concentration of enzyme, C_{E_0}, but once the reaction is under way the free enzyme concentration will be less than this as some has been converted to complex, *ES*. At any instance the total amount of free and bound enzyme is equal to C_{E_0}, so

$$C_{E_0} = C_{ES} + C_E$$

and
$$C_{ES} = C_{E_0} - C_E = \frac{(k_1\,C_E\,C_S)}{(k_{-1} + k_2)}$$

Hence,
$$C_{E_0} = C_E + \frac{(k_1\,C_E\,C_S)}{(k_{-1} + k_2)}$$

or
$$C_{E_0} = C_E\left[1 + \frac{(k_1\,C_S)}{(k_{-1} + k_2)}\right]$$

or
$$C_E = \frac{C_{E_0}}{\left[1 + \frac{(k_1\,C_S)}{(k_{-1} + k_2)}\right]}$$

or
$$C_E = \frac{C_{E_0}}{\left(1 + \frac{C_S}{k_S}\right)}$$

where,
$$k_S = \frac{(k_{-1} + k_2)}{k_1}$$ and is called **Michaelis Constant**.

Thus, the rate of enzyme-catalysed reaction is

$$\text{Rate} = k_2 \left(\frac{C_S}{k_S}\right)\left(\frac{C_{E_0}}{\left(1 + \frac{C_S}{k_S}\right)}\right)$$

$$\text{Rate} = \frac{k_2\, C_{E_0}\, C_S}{(k_S + C_S)}$$

This relation is known as **Michaelis–Menton Law**.

If k_S is small as compared to C_S, *i.e.*, all the enzyme is present in the form of complex, the rate will be maximum (R_{max}),

$$R_{max} = k_2\ C_{E_0}$$

Thus maximum rate is independent of substrate concentration. Hence, rate at instant (R_t),

$$R_t = \frac{R_{max}\, C_S}{(k_S + C_S)}$$

or
$$\left(\frac{1}{R_t}\right) = \left(\frac{1}{R_{max}}\right) + \frac{k_S}{(R_{max}\ C_S)}$$

Thus a plot of reciprocal of reaction velocity at any time vs reciprocal of substrate concentration would be a straight line, whose slope will be equal to $\frac{k_S}{(R_{max})}$ and intercept from *y*-axis will be $\left(\frac{1}{R_{max}}\right)$. These results can be used to get values of k_S and R_{max}.

The Michaelis Constant (k_S) is characteristic of each enzyme and independent of substrate as well as enzyme concentration. We can write

$$k_S = \frac{(R_{max} - R_t)\, C_S}{R_t}$$

Hence, if $$R_t = \frac{(R_{max})}{2} \; ; k_S = C_S$$

Thus, Michaelis constant (k_S) is the substrate concentration at half the maximum reaction rate.

Autocatalysis

In certain reactions the product formed during the reaction catalyses the original reaction. Therefore, the rate of reaction becomes faster as the reaction proceeds. Such reactions are autocatalysed and this type of catalysis is called autocatalysis. Few examples of autocatalysis are given below :

(*i*) Oxidation of oxalic acid by acidic permanganate is catalysed by manganous ion formed in the reaction.

$2KMnO_4 + 6HCl + 5\,(COOH)_2 \longrightarrow 2MnCl_2 + 10CO_2 + 8H_2O + 2KCl$ Thus, when $KMnO_4$ is added to oxalic acid the colour of $KMnO_4$ is decolourized slowly but as the reaction proceeds Mn^{2+} ion is formed which catalyses the reaction. As a result the colour of $KMnO_4$ is decolourized rapidly.

(*ii*) In acidic hydrolysis of ester, the product is a fatty acid which catalyses the reaction

$$CH_3COOC_2H_5 + H_2O \xrightarrow{H^{\oplus}} CH_3COOH + C_2H_5OH.$$

Hence the reaction is slow initially but as more CH_3COOH is formed reaction rate increases.

(*iii*) In reaction of nitric acid on copper, the reaction is catalysed by one of the product of the reaction, *i.e.*, nitrous oxide. Hence, evolution of gases is slow in the beginning but as nitrous oxide is formed the rate of evolution of gases increases.

$$4Cu + 2NO_3^{\ominus} + 10H^{\oplus} \longrightarrow 4Cu^{2\oplus} + N_2O + 5H_2O$$

Induced Catalysis

If the presence of one reaction influences the speed of other reaction, the phenomenon is known as induced catalysis.

For example, oxidation of sodium arsenite in solution by passing air is an extremely slow reaction. On the other hand, oxidation of sodium sulphite solution by passing air is fairly fast.

When air is passed through a solution containing both sodium arsenite as well as sodium sulphite, the oxidation of both occurs fairly fast. Thus, oxidation of sodium sulphite catalyzes the oxidation of sodium arsenite.

Isopropyl alcohol is not ordinarily oxidized by air but in the presence of sodium sulphite, it is oxidized to acetone and sodium sulphite to sulphate.

Negative Catalysis

A substance that decreases the rate of a reaction is called a negative catalyst. These negative catalysts find important role in industry as they are used to control the rate of reaction and for stabilizing unstable products. Examples are as follows :

Small amounts of acetanilide are added to hydrogen peroxide in order to decrease its rate of decomposition. In the oxidation of sodium sulphite by oxygen in solution mannitol acts as a negative catalyst.

Addition of tetraethyl lead to petrol retards its combustion.

5.13 MISCELLANEOUS EXAMPLES

5.13.1 Monsanto Process for the Production of Acetic acid

The Monsanto process is an important method for the manufacture of acetic acid. This process operates at a pressure of 30–60 atm and a temperature of 150–200°C and gives selectivity greater than 99%.

The catalyst employed is the anion *cis*-[[Rh $(CO)_2$ I_2]]$^-$. (1) The catalytic cycle involves six steps, two of which do not involve rhodium : conversion of methanol to methyl iodide and the hydrolysis of the acetyl iodide to acetic acid. The first organometallic step is the oxidative addition of methyl iodide to *cis*-[[Rh $(CO)_2$ I_2]]$^-$ to form the hexacoordinate species [[(CH_3) Rh $(CO)_2$ I_3]]$^-$. (2) This anion rapidly transforms, *via* the migration of a methyl group to the carbonyl ligand, affording the pentacoordinate acetyl complex [[(CH_3CO) Rh (CO) I_3]]$^-$. (3) This five coordinate complex then reacts with carbon monoxide to form the six coordinate dicarbonyl complex. (4) This decomposes by reductive elimination to form acetyl iodide (CH_3COI) and regenerates the active form of the catalyst. Acetyl iodide is then hydrolysed to acetic acid.

$$\underset{(1)}{CH_3I + cis\text{-}[Rh\,(CO)_2\,I_2]^-} \xrightarrow[\text{addition}]{\text{Oxidative}} \underset{(2)}{[(CH_3)\,Rh\,(CO)_2\,I_3]^-}$$

$$\xrightarrow{\text{migration}} \underset{(3)}{[(CH_3CO)\,Rh\,(CO)\,I_3]^-} \xrightarrow[\text{addition}]{\text{CO}} \underset{(4)}{[(CH_3CO)\,Rh\,(CO)_2\,I_3]^-}$$

$$\xrightarrow[\text{elimination}]{\text{reductive}} (1) + CH_3COI \xrightarrow{H_2O} CH_3COOH$$

5.13.2 Ziegler–Natta Catalyst

A Ziegler-Natta catalyst is a mixture of reagents used in the production of polymers of 1-alkenes (α-olefins). Ziegler–Natta catalysts are based on titanium and organometallic aluminium compounds, for example triethylaluminium, $(C_2H_5)_3$ Al.

Ziegler–Natta catalysts are used to polymerize terminal 1-alkenes.

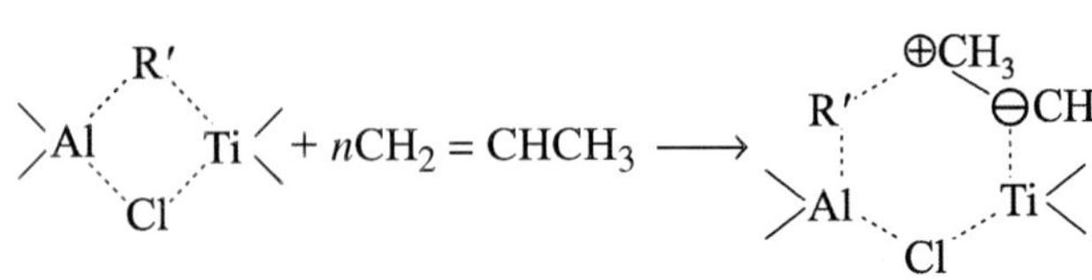

The first Ziegler-Natta catalyst was produced by treating crystalline α-$TiCl_3$ with $[AlCl(C_2H_5)_2]_2$. Polymerization reactions of any alkene occurs at special Ti centers located on the exterior of the crystallites. Most titanium ions in these crystallites are surrounded by six chloride ligands to give an octahedral structure. At the surface, however, "defects" occur where some Ti centers lack their full complement of chloride ligands. The alkene molecule binds at these "vacancies". The alkene converts to an alkyl ligand group. The most probable pathway of this reaction is the insertion of the $C = C$ bond of the alkene molecule into the Ti-C bond:

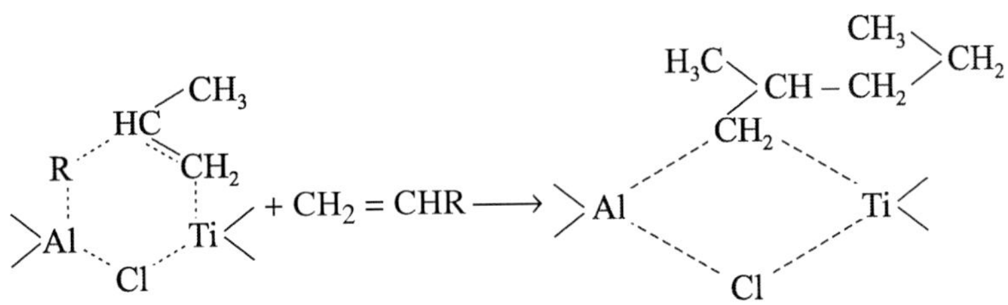

The coordination sphere of the Ti atom restricts the approach of incoming alkene molecules, thereby imposing stereoregularity on the growing polymer chain.

5.13.3 Hoechst–Wacker Process

This process refers to the oxidation of ethylene to acetaldehyde by oxygen in the presence of a palladium tetrachloride catalyst.

The catalytic process be described as :

$$[PdCl_4]^{2-} + C_2H_4 + H_2O \longrightarrow CH_3CHO + Pd + 2HCl + 2Cl^-$$

$$Pd + 2CuCl_2 + 2Cl^- \longrightarrow [PdCl_4]^{2-} + 2CuCl$$

$$2CuCl + \frac{1}{2}O_2 + 2HCl \longrightarrow 2CuCl_2 + H_2O$$

5.13.4 Heck Reaction

The Heck reaction (also called the Mizoroki-Heck reaction) is the chemical reaction of an unsaturated halide with an alkene and a strong base and palladium catalyst to form a substituted alkene.

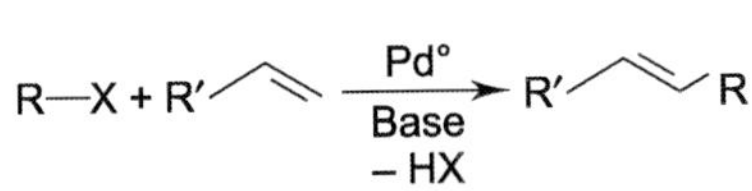

The reaction is performed in the presence of an organopalladium catalyst. The halide is an aryl, benzyl, or vinyl compound and the alkene contains at least one proton and is often electron-deficient such as acrylate ester or an acrylonitrile. The catalyst can be tetrakis (triphenylphosphine) palladium (0).

The ligand is triphenylphosphine. Base is triethylamine, potassium carbonate or sodium acetate.

The Heck reaction is applied industrially in the production of **naproxen** and the **sunscreen**.

Objective Type Questions

1. The rate constant of a reaction has same units as the rate of reaction. The reaction is of
   (*a*) Zero order (*b*) First order
   (*c*) Second order (*d*) None of these.

2. If a is the initial concentration of reactant, then the half-life period of a reaction of nth order will be
   (*a*) a^{n-1} (*b*) a^{1-n}
   (*c*) a^{n+1} (*d*) a^n.

3. For a reaction a plot of log $(a - x)$ vs time (t) is a straight line with slope equal to $\frac{k}{2.303}$, the reaction is of
   (*a*) Zero order (*b*) First order
   (*c*) Second order (*d*) Third order.

4. Which of the following statement is false?
   (*a*) Order of reaction can be fractional
   (*b*) Reaction of higher order are rare
   (*c*) Order is same as molecularity of the reaction
   (*d*) Rate of reaction is proportional to specific rate constant.

5. Consider a reaction of first order $P \rightarrow Q$. Initial concentration of P is C_0 while at time t the concentration of Q is C_x, then rate equation may be written as
   (*a*) $K = \frac{1}{t}\log_e\frac{(C_0 - C_x)}{C_0}$ (*b*) $K = \frac{1}{t}\log_e(C_0 - C_x)$
   (*c*) $K = \frac{2.303}{t}\log_e\frac{(C_0 + C_x)}{(C_0 - C_x)}$ (*d*) $K = \frac{1}{t}\log_e\frac{C_0}{(C_0 - C_x)}$

6. The equation for the rate constant is given by $K = pZ\ \exp^{-\frac{E_a}{RT}}$. Then a chemical reaction will proceed more rapidly if there is a decrease in
   (*a*) T (*b*) Z
   (*c*) E_a (*d*) p.

7. The rate of a second order reaction is 3×10^{-5} mol l^{-1} s^{-1} when the initial concentration is 0.2 mol l^{-1}. The rate constant for the reaction will be
   (*a*) 7.5×10^{-4} mol l^{-1} s^{-1} (*b*) 1.5×10^{-3} mol^{-1} l s^{-1}
   (*c*) 1.5×10^{-4} mol l^{-1} s^{-1} (*d*) 7.5×10^{-4} mol^{-1} l s^{-1}.

8. A catalyst is a substance which

(*a*) Changes the equilibrium constant of the reaction

(*b*) Brings equilibrium sooner

(*c*) Increases the equilibrium concentration of products

(*d*) Initiates a reaction.

9. Michaelis-Menton law is related to

(*a*) Acid-base catalysis (*b*) Heterogeneous catalysis

(*c*) Enzyme catalysis (*d*) Homogeneous catalysis.

10. Half-life period of a first order reaction is 69.3 seconds. Its rate constant is

(*a*) $10^{-4}\ s^{-1}$ (*b*) $10^{-2}\ s^{-1}$

(*c*) $10\ s^{-1}$ (*d*) $10^{2}\ s^{-1}$.

11. The rate of reaction between A and B increases by a factor of 100 when the concentration of A is increased ten times. The order of reaction with respect to A is

(*a*) Zero (*b*) Half

(*c*) One (*d*) Two.

12. In the catalytic conversion of N_2 to NH_3 by Haber's process the rate of formation of NH_3 is 4×10^{-3} mol $l^{-1}\ s^{-1}$. What would be the rate of disappearance of H_2

(*a*) 1×10^{-3} mol $l^{-1}\ s^{-1}$ (*b*) 2×10^{-3} mol $l^{-1}\ s^{-1}$

(*c*) 4×10^{-3} mol $l^{-1}\ s^{-1}$ (*c*) 6×10^{-3} mol $l^{-1}\ s^{-1}$.

13. In the reversible reaction $2NO_2 \underset{k_2}{\overset{k_1}{\rightleftharpoons}} N_2O_4$, the rate of disappearance of NO_2 is equal to

(*a*) $\frac{2k_1}{k_2}[NO_2]^2$ (*b*) $2k_1\ [NO_2]^2 - 2k_2\ [N_2O_4]$

(*c*) $2k_1\ [NO_2]^2 - k_2\ [N_2O_4]$ (*d*) $[2k_1 - k_2]\ [NO_2]$.

14. In the sequence of reaction $P \xrightarrow{k_1} Q \xrightarrow{k_2} R \xrightarrow{k_3} S$ if $k_3 > k_2 > k_1$, then rate determining step is

(*a*) $P \rightarrow Q$ (*b*) $Q \rightarrow R$

(*c*) $R \rightarrow S$ (*d*) $P \rightarrow S$.

15. The reaction $2O_3 \longrightarrow 3O_2$ proceeds by the following mechanism :

$$O_3 \longrightarrow O_2 + O \text{ (Fast)} ;\ O_3 + O \longrightarrow 2O_2 \text{ (Slow)}$$

The rate law expression should be

(*a*) Rate = $k\ [O_3]^2$ (*b*) Rate = $k\ [O_3]\ [O_2]$

(*c*) Rate = $k\ [O_3]\ [O_2]^{-1}$ (*d*) Rate = $k\ [O_3]^2\ [O_2]^{-1}$.

ANSWERS

1. (*a*)	**2.** (*b*)	**3.** (*b*)	**4.** (*c*)	**5.** (*d*)
6. (*c*)	**7.** (*d*)	**8.** (*b*)	**9.** (*c*)	**10.** (*b*)
11. (*d*)	**12.** (*d*)	**13.** (*c*)	**14.** (*a*)	**15.** (*d*).

Short Answer Type Questions

1. Distinguish between rate and rate constant.

2. Comment on the statement "Higher the activation energy slower is the rate of reaction".

3. Distinguish between molecularity and order of reaction.

4. Write an expression for the units of rate constant for n^{th} order reaction.

5. Explain the following terms in brief :

(*a*) Half-life period (*b*) Pseudo order reaction.

6. What is autocatalysis ? Give one example.

7. Give one example each of

(*a*) Negative catalyst (*b*) Induced catalysis

8. Name three important factors that influence rate of reaction.

9. How can the age of a fossil be determined ?

10. Give one example of parallel reaction.

Subjective Questions

1. Derive the rate equation for a first order reaction and show that $T_{1/2}$ for this is equal to $0.693/K$.

2. Discuss important methods for the determination of order of reaction.

3. Give the effect of temperature on reaction rate constant and derive Arrhenius equation.

4. What is transition state theory of reaction rate ? Give its thermodynamical treatment.

5. Discuss the kinetics of enzyme catalysis.

6. From the following data for the reaction $2A + 2B \longrightarrow$ product find the order of reaction:

$[A]$	0.1	0.1	0.2	0.4
$[B]$	0.2	0.4	0.4	0.2
Rate mol l^{-1} s^{-1}	2×10^{-3}	4×10^{-3}	4×10^{-3}	2×10^{-3}

[**Ans.** Order is 1 with respect to B and zero with respect to A]

7. The half-life periods for a reaction at initial concentration of 1.0 and 2.0 mol l^{-1} are 1 min. 40 s and 50 s, respectively. What should be the order of reaction ?
[**Ans.** Second order]

8. How many times the rate of reaction change if the temperature of a reaction is changed from 300 to 400 K ? Given that the activation energy of the reaction is 42.0 kJ

[**Ans.** 67.3 times]

9. Half-life for the decomposition of ^{238}U is 4.5×10^9 years. How many atoms of it disintegrate in twenty four hours ?

[**Ans.** 1×10^9 atoms]

10. In decomposition reaction of N_2O_5 it is observed that specific rate constant for the reaction is 3.5×10^{-5} and 4.2×10^{-4} at 300 and 320 K, respectively. Calculate activation energy for the reaction.

[**Ans.** 99 kJ]

11. Following reaction was found to be second order

$$CH_3CH_2CH_2Br + S_2O_3^{2\ominus} \longrightarrow CH_3CH_2CH_2S_2O_3^{\ominus} + Br^{\ominus}$$

The reaction was studied by titrating free thiosulphate ion with iodine solution and following results were obtained

Time (s)	0	1100	2000	3000
Concentration of S_2O_3 (mol l^{-1})	0.1	0.09	0.086	0.081

[**Ans.** $K = 1.6 \times 10^{-3}$ mol^{-1} l s^{-1}]

12. The optical rotation of sucrose in 0.1 N HCl at various time intervals is given below:

Time (s)	0	600	1200	∞
Sp. rotation	+ 32.4	+ 28.7	+ 25.6	– 11.0

Show that the reaction is first order and calculate specific rate constant.

[**Ans.** $K = 8.6 \times 10^{-3}$ s^{-1}]

13. The hydrolysis of ethyl benzoate by aqueous sodium hydroxide taken in equal concentrations was studied by titrating against 1.0 N acid at different stages of reaction. The results are given below :

Time (s)	0	140	330	800
mL of acid used	10.0	9.0	8.0	6.0

Show that the reaction is second order and calculate specific rate constant.

[**Ans.** $K = 7.93 \times 10^{-2}$ mol^{-1} l s^{-1}]

14. In the reaction $2FeCl_3 + SnCl_2 \longrightarrow SnCl_4 + 2FeCl_2$ using equal concentrations of $FeCl_3$ and $SnCl_2$ following results were obtained:

Time (s)	0	120	300	600
Concentration of $FeCl_3$	0.14	0.084	0.061	0.045

Show that the reaction is third order and calculate rate constant.

[**Ans.** $K = 8.84 \times 10^{1}$ mol^{-2} l^{2} s^{-1}]

15. Calculate the number of desintegration per second produced by 1 g of radium. $T_{1/2}$ of ^{226}Ra is 1600 years.

[**Ans.** 3.7×10^{10} dps]

16. Wooden sample from a fossil tree gave 10.78 counts per minute per gram of carbon due to ^{14}C whereas a freshly cut wooden sample gives 15.28 counts per minute per gram of carbon. Calculate the approximate age of fossil. $T_{1/2}$ of ^{14}C is 5600 years.

[**Ans.** 2890 years]

17. Analysis of a sample of uranium ore showed it to contain 20.6% ^{206}Pb and rest ^{238}U. Half–life of ^{238}U is 4.51×10^{9} years. It is assumed that all the lead has come from the decay of uranium, calculate the age of ore.

[**Ans.** 1.22×10^{9} years]

❑❑❑